Systems Engineering

People want to create a better world and planet; however, where, and how to start remains the question. Systems Engineering's problem-solving methodology can help with its ability to answer multiple questions along with connecting actions and impacts. This book uses the Systems Engineering problem-solving methodology to frame how each answer impacts the planet when multiple actions are strung together no matter where they take place.

Systems Engineering: Influencing Our Planet and Reengineering Our Actions illustrates a hierarchical Systems Engineering view of the world with each individual in mind as a link in the chain. It uses an Industrial Engineering framework for action implementations and identifies humans' interconnected actions. The book discusses the implementation of the Systems Engineering problem-solving methodology and leverages existing concepts of environmental sustainability. A template is present for personal actions for environment social responsibility using a Systems Engineering problem-solving approach and focuses on the foundational use of the trademarked DEJI Systems Model® for action design, evaluation, justification, and integration.

This book is a perfect read for all academic disciplines and all engineering fields, as well as business and management fields. It reminds us of the Environmental Foundation of NAE's 14 Grand Challenges and the part we can play.

Systems Innovation Book Series
Series Editor: Adedeji B. Badiru

Systems Innovation refers to all aspects of developing and deploying new technology, methodology, techniques, and best practices in advancing industrial production and economic development. This entails such topics as product design and development, entrepreneurship, global trade, environmental consciousness, operations and logistics, introduction and management of technology, collaborative system design, and product commercialization. Industrial innovation suggests breaking away from the traditional approaches to industrial production. It encourages the marriage of systems science, management principles, and technology implementation. Particular focus will be the impact of modern technology on industrial development and industrialization approaches, particularly for developing economics. The series will also cover how emerging technologies and entrepreneurship are essential for economic development and society advancement.

Global Supply Chain
Using Systems Engineering Strategies to Respond to Disruptions
Adedeji B. Badiru

Systems Engineering Using the DEJI Systems Model®
Evaluation, Justification, Integration with Case Studies and Applications
Adedeji B. Badiru

Handbook of Scholarly Publications from the Air Force Institute of Technology (AFIT), Volume 1, 2000–2020
Edited by Adedeji B. Badiru, Frank Ciarallo, and Eric Mbonimpa

Project Management for Scholarly Researchers
Systems, Innovation, and Technologies
Adedeji B. Badiru

Industrial Engineering in Systems Design
Guidelines, Practical Examples, Tools, and Techniques
Brian Peacock and Adedeji B. Badiru

Leadership Matters
An Industrial Engineering Framework for Developing and Sustaining Industry
Adedeji B. Badiru and Melinda Tourangeau

Systems Engineering
Influencing Our Planet and Reengineering Our Actions
Adedeji B. Badiru

Systems Engineering
Influencing Our Planet and Reengineering Our Actions

Adedeji B. Badiru

CRC Press
Taylor & Francis Group
Boca Raton London New York

CRC Press is an imprint of the
Taylor & Francis Group, an **informa** business

Designed cover image: Shutterstock

First edition published 2024
by CRC Press
2385 Executive Center Drive, Suite 320, Boca Raton, FL 33431

and by CRC Press
4 Park Square, Milton Park, Abingdon, Oxon, OX14 4RN

CRC Press is an imprint of Taylor & Francis Group, LLC

© 2024 Adedeji B. Badiru

ISBN: 978-1-032-24510-2 (hbk)
ISBN: 978-1-032-24511-9 (pbk)
ISBN: 978-1-003-27905-1 (ebk)

DOI: 10.1201/9781003279051

Typeset in Times
by SPi Technologies India Pvt Ltd (Straive)

Dedication

Dedicated to the essence of **"Aloha Aina,"** which is a Hawaiian concept of stewardship of the land and natural resources around us, with environmental commitment that is deeply rooted in the Hawaiian culture and tradition. In the Hawaiian native language, Aina refers to the **land**, which feeds and nourishes humans. If we influence Aina positively, in the premise of this book, the land (i.e., our planet) will continue to take care of us and preserve existence.

Contents

Preface

It is a systems world.

Things often start gradually. One thing leads to another and another. Before we know it, the whole system is affected. We need to reengineer our action. It is important to reemphasize that the premise of this new book is to put the onus on individuals with respect to reengineering personal actions that can contribute to our collective influence on the planet. Using a systems perspective, the book will get individuals excited and committed to doing their respective parts in the end goal of influencing the planet positively. Note that the book did not say "saving our planet," as most sensational headlines might say. But, rather, it focuses on what individuals can do to contribute to the overall good influence.

The reason for this book is based on the call for urgent actions across all socio-economic platforms of the world on how we can combat global warming and collectively move toward saving the planet from an environmental disaster. "No action" is not acceptable. But what actions are needed where, when, where, why, and how?

The rapid decline of the environment is alarming. It is no longer in doubt that climate change is occurring. Scientific studies around the world have confirmed how human actions collectively affect the environment, the climate, and the weather, which reciprocally affects each and every one of us. In this regard, we all have the responsibility to pattern our actions in ways that will be beneficial rather than adverse to the planet in the long run. This book outlines the what, who, when, where, why, and how of human impacts on the environment. Throughout history, engineers, along with other STEM professionals, have stepped forward to tackle pressing issues of the world. Now, more than ever, engineers are needed to help combat the looming environmental disaster caused by climate change. The book uses the problem-solving framework of systems engineering and the human-centric principles of industrial engineering. None of us is immune to the effects of climate change, which is causing more frequent and more intense weather phenomena. It is through a structured systems approach that individuals can be convinced of how each action counts. The book's contents include Impacts of Humans on the planet, Principles of Industrial engineering for the environment, Environmental foundation of NAE's 14 Grand Challenges, Process improvement tools for managing the environment, Engineering's role in the society of the past, present, and future, How people make up the environment, Influence diagram of global environmental flow, and Environment case examples. Written from a practical standpoint, this book will get readers knowledgeable and excited about what is needed from each of us to participate in the urgent environment remedial actions needed to save the planet. Of great focus in the book is the foundational use of the trademarked DEJI Systems Model® for action design, evaluation, justification, and integration. Integration is the key in mapping our individual and collective actions to the desired end results of influencing our planet positively.

Acknowledgements

I acknowledge and appreciate the inspirational voices of those around the world, who continue to commit themselves and their resources so that we can collectively influence our planet and our environment positively. At the personal level, I thank all the personnel of CRC Press and Taylor & Francis Group for their unflinching embrace and support of the wide publishing topics in my Systems Innovation Book series. Of particular note is the recognition of Ms. Cindy Carelli and her extraordinary editorial team for their indefatigable commitment to the growth of the book series, with practical guidance on addressing critical issues that advance our foray into new ideas in science, technology, engineering, and mathematics. I also thank my international multidisciplinary research collaborators in Germany, Dr. Nils Wagenknecht and Mr. Andreas Mertens, for their intellectual contributions to the development of this book manuscript.

About the Author

Adedeji Badiru is a professor of Systems Engineering at the Air Force Institute of Technology (AFIT). He was previously professor and head of Systems Engineering and Management at AFIT, professor and department head of Industrial Engineering at the University of Tennessee—Knoxville, and professor of Industrial Engineering and dean of University College at the University of Oklahoma, Norman. He is a registered professional engineer (PE), a certified project management professional (PMP), a fellow of the Institute of Industrial & Systems Engineers, a fellow of the Industrial Engineering and Operations Management Society, and a fellow of the Nigerian Academy of Engineering. He is also a program evaluator for ABET. He holds a leadership certificate from the University Tennessee Leadership Institute. He has a BS in Industrial Engineering, an MS in Mathematics, and an MS in Industrial Engineering from Tennessee Technological University, and a PhD in Industrial Engineering from the University of Central Florida. His areas of interest include mathematical modeling, project modeling and analysis, economic analysis, systems engineering modeling, computer simulation, and productivity analysis. He is a prolific author, with over 38 books, over 32 book chapters, over 140 journal and magazine articles, and over 200 conference presentations. He is a member of several professional associations and scholastic honor societies. Fittingly for the theme of this book, Adedeji is an active member of the Beavercreek Environmental Advisory Committee (BEAC) in the city of Beavercreek, Ohio, USA.

Adedeji Badiru, a world-renowned educator, has won several awards for his teaching, research, administrative, and professional accomplishments. Some of his selected awards include the 2009 Dayton Affiliate Society Council Award for Outstanding Scientists and Engineers in the Education category with a commendation from the 128th Senate of Ohio, 2010 ASEE John Imhoff Award for his global contributions to Industrial Engineering Education, the 2011 Federal Employee of the Year Award in the Managerial Category from the International Public Management Association, Wright Patterson Air Force Base, the 2012 Distinguished Engineering Alum Award from the University of Central Florida, the 2012 Medallion Award from the Institute of Industrial Engineers for his global contributions in the advancement of the profession, 2016 Outstanding Global Engineering Education Award from the Industrial Engineering and Operations Management, 2015 Air Force-level winner of the National Public Service Award from The American Society for Public Administration and the National Academy of Public Administration, 2013 Father D. J. Slattery Excellence Award from Saint Finbarr's College Alumni Association – North America, 2013 Award Team Leader, Air Force Organizational Excellence Award for Air University C3 (Cost Conscious Culture), 2013 Finalist for Jefferson Science Fellows Program, National Academy of Sciences, 2012 Book-of-the-Month Recognition for *Statistical Techniques for Project Control* from the Industrial Engineering Magazine, the 2009 Industrial Engineering Joint Publishers Book-of-the-Year Award for *The Handbook of Military Industrial Engineering* and 2020 for *The Story of Industrial Engineering*.

Adedeji is also the book series editor for CRC Press/Taylor & Francis Group book series on Systems Innovation. He has served as a consultant to several organizations around the world, including Russia, the UK, Canada, Mexico, Taiwan, Nigeria, and Ghana. He has conducted customized training workshops for numerous organizations, including Sony, AT&T, Seagate Technology, the US Air Force, Oklahoma Gas & Electric, Oklahoma Asphalt Pavement Association, Hitachi, the Nigeria National Petroleum Corporation, and ExxonMobil. He has served as a technical project reviewer, curriculum reviewer, and proposal reviewer for several organizations, including The Third-World Network of Scientific Organizations, Italy, the Social Sciences and Humanities Research Council of Canada, the National Science Foundation, the National Research Council, and the American Council on Education. He is on the editorial and review boards of several technical journals and book publishers. Prof. Badiru has also served as an Industrial Development Consultant to the United Nations Development Program. In 2011, Prof. Badiru led a research team to develop analytical models for Systems Engineering Research Efficiency (SEER) for the Air Force acquisitions integration office at the Pentagon. He has led a multi-year multi-million-dollar research collaboration between the Air Force Institute of Technology and KBR Aerospace Group. Prof. Badiru has diverse areas of avocation. His professional accomplishments are coupled with his passion for writing about everyday events and interpersonal issues, especially those dealing with social responsibility. Outside of the academic realm, he writes motivational poems, editorials, and newspaper commentaries, as well as engaging in paintings and crafts. Adedeji is the 2020 recipient of the Lifetime Achievement Award from Taylor & Francis Group. He was also part of the AFIT team that led the institution's receipt of the 2019/2020 US Air Force Organizational Excellence Award. He is also the recipient of the 2022 BEYA Career Achievement Award in Government category. He holds a US trademark for the DEJI Systems Model for design, evaluation, justification, and integration.

1 Systems Engineering View of Climate Change

SYSTEMS BACKDROP FOR CLIMATE CHANGE

In NAS & TRS (2020), the US National Academy of Sciences and the British Royal Society provide an instructive setting for how we must respond to climate change from a systems perspective. The joint report affirms that climate change is one of the defining issues of our time. There is no doubt about this, based on recent experiences with extreme weather patterns. According to the NAS & TRS report, it is now more obvious and certain than ever before, based on many lines of evidence, that humans are changing Earth's climate. The atmosphere and oceans have warmed, which has been accompanied by sea level rise, a noticeable decline in Arctic sea ice, and related catastrophic weather patterns. The impacts of climate change on people and nature are increasingly observed. Unprecedented flooding, heatwaves, and wildfires have cost billions of dollars in damages. Human and animal habitats are undergoing rapid shifts in response to changing temperatures and precipitation patterns. It is a sorry and discouraging narrative all across the planet (see NRC, 2010 and NAS, 2018). There is only one planet and we are all interconnected residents of the planet. We must all work together to protect the planet and make it livable for everyone. Whether we love our neighbors or not, we must all remain on the same planet together. What affects one affects all. If we all work together, from a systems perspective, we can build a more sustainable planet that can facilitate a healthier future for all. Let's join the journey, as enjoined by Badiru and Agustiady (2021).

Whether we see it or not, we are in a rapidly changing world. Our actions are essential and significant as impacts on the world system. Every little action can go a long way, but only if we all realize what actions can help. This book is about the what, who, when, where, how, and why of influencing our planet positively.

Men and women form the core of any system. This is in alignment with the recognition that organizations operate and thrive on the efficient interplay of people, process, and tools. The first object of any good system is the people. Thus, developing first-class people in the workforce must be a priority of any organization or group making up a work system. From a systems perspective, every little bit contributed by a whole lot of people can amount to a monumental accomplishment, as reiterated by the following famous quotes:

In the past, the man has been first; in the future, the system must be first.
— **Frederick Winslow Taylor** (1856–1915)

Nothing is particularly hard if you divide it into small jobs.
— **Henry Ford** (1863–1947)

DOI: 10.1201/9781003279051-1

It is a systems world. We must treat it as such in order to influence it positively. Things often start gradually. One thing leads to another and another. Before we know it, the whole system is affected. We need to re-engineer our actions. The problem of climate change has been creeping on us for years. Although a potpourri of concerns, actions, and pledges has been floating around for years, everything came to a head in 2021 when we suddenly realized the immense changes in the climate, manifested in more frequent and more extreme weather patterns. A recent USA national weather warning says:

> Temperatures across the country will hit record highs and lows as a huge winter storm sweeps across much of the country. In D.C., temperatures will reach nearly 80 degrees before plummeting to below freezing. The National Weather Service says the storm will be disruptive to travel, infrastructure, livestock, and recreation in affected areas.

The premise of this book is that there is something each one of us can do at our own elementary level to influence our planet positively. Unfortunately, we, as individuals, don't always realize the extent that our actions can go in impacting the planet, either positively or negatively. As a case example, I once witnessed a simple action that caused a huge impact, but to the nonchalance of the actor. I was stopped at a red light. A pickup truck in the adjacent lane sped up furiously across the intersection in the attempt to beat the red light. He made it. But at what expense? A huge black smoke spewed out from the pickup's exhaust and drifted into the atmosphere. The driver jubilated happily for beating the light. He was happy. But the environment was sad because of, yet another, human assault. It is my belief, as an author, that the more we know what is expected of us and the more we know the implications, and the more likely we are to join the global efforts to save our planet. Badiru (2016) raised an alarm of climate change from a Systems Engineering framework. Some of his arguments center on several factors, including the following:

* Water, land, and air all interface rhythmically to create the environment, in which humans live and work.
* Climate is the precursor to all human activities.
* We must take a systems view of the world to respond to climate change issues.

ENERGY EVOLUTION AND REVOLUTION

There is no shortage of papers, journal articles, seminars, presentations, statements, communiqués, conventions, conferences, and press releases on the urgency of taking action on the prevailing environmental issues on our planet. The problem has not been neglected, yet it does not appear that we are making progress in remediating the damage already done and stemming future damages. Jones (2014) presents a warning that the energy landscape of the future will not be a sudden break from the past.

Organizations will have to evolve their portfolios and action along the paths of energy revolution. He provides the following heads-up on what the world needs to do:

- Seek out alternatives
- Leverage the local advantage
- Find the way that works
- Make smart investments
- Expect the unexpected
- Adapt to the environment
- Learn from the past

It is the position of this book that doing what we need to do can best be approached from a global systems framework. In a related case study, Gamble (2014) presents how Australia is taking the lead in many environmental strategies. His conclusion highlights how an Australian project proved that renewable energies can play more than a supporting role in addressing global climate change dilemmas.

THE MARCH TOWARDS RENEWABLES

Sambo (2023) emphasizes that reasonable standard of living and economic growth both require adequate, reliable, and affordable energy supplies for households, business services, commercial enterprises, transportation systems, and industrial sectors of all national economies. Fossil-type energy resources, like oil and natural gas and also coal, which have been the dominant sources of energy for electrification, transport and industrial production, are highly polluting and exhaustible. 87% of CO_2 emissions from human sources arise from the combustion of fossil fuels and this has four major components:

- **Electricity generation: 41%**
- **Transportation: 23%**
- **Industry: 20%**
- **Others: 16%**

Renewable energy (RE) sources, such as solar, wind, hydro, geothermal, biomass, and others, are, however, in principle non-exhaustible as they are always available on a cyclic or periodic basis and they emit only negligible amounts of greenhouse gases. There is an urgent need to significantly minimize carbon dioxide (CO_2) emissions in the power and transportation sectors. Sustainable pathways should be adopted for the stability of the environment and remediating impacts on climate change. It is noted that in 2015, the United Nations adopted the 17 Sustainable Development Goals (SDGs) to, among other things, by 2030, curtail the catastrophe that will befall the world from the increasing menace of global warming and climate change. The 17 SDGs are:

Goal 1: End poverty in all its forms everywhere
Goal 2: End hunger, achieve food security and improved nutrition, and promote sustainable agriculture

Goal 3: Ensure healthy lives and promote well-being for all at all ages
Goal 4: Ensure inclusive and equitable quality education and promote lifelong learning opportunities for all
Goal 5: Achieve gender equality and empower all women and girls
Goal 6: Ensure availability and sustainable management of water and sanitation for all
Goal 7: Ensure access to affordable, reliable, sustainable and modern energy for all
Goal 8: Promote sustained, inclusive and sustainable economic growth, full and productive employment and decent work for all
Goal 9: Build resilient infrastructure, promote inclusive and sustainable industrialization and foster innovation
Goal 10: Reduce inequality within and among countries
Goal 11: Make cities and human settlements inclusive, safe, resilient and sustainable
Goal 12: Ensure sustainable consumption and production patterns
Goal 13: Take urgent action to combat climate change and its impacts
Goal 14: Conserve and sustainably use the oceans, seas and marine resources for sustainable development
Goal 15: Protect, restore and promote sustainable use of terrestrial ecosystems, sustainably manage forests, combat desertification, and halt and reverse land degradation and halt biodiversity loss
Goal 16: Promote peaceful and inclusive societies for sustainable development, provide access to justice for all and build effective, accountable and inclusive institutions at all levels
Goal 17: Strengthen the means of implementation and revitalize the global partnership for sustainable development

The best way to achieve most or all the above goals is to use a comprehensive systems-based approach, specifically, leveraging the efficacy of systems engineering. This is where, why, and when individual contributions are essential, primarily through our energy consumption practices. Badiru and Osisanya (2013) present the systems-based interrelationships of energy generation, transmission, distribution, and consumption as summarized below:

- Exploration: Drilling, fracking, dissolution removal technology
- Generation: Fossil, fission, fusion
- Distribution: Transmission technology, wireless transfer, storage movement
- Consumption: Households, industry, transportation, business, construction

In tackling the climate change problems, both qualitative and quantitative approaches must be pursued. This allows us to bring in technological approaches (Lathan, 2022) as well as human-centered social considerations. Again, the premise of this book is that a systems engineering approach is key to making the desired progress. As such, this book contains a mix of technical, social, economic, engineering, and political considerations.

THE HYDROGEN ANGLE

One of the emerging exciting technical approaches to the future of energy on Earth is hydrogen-based technologies. A 2023 article (Mines Staff, 2023) on the ongoing research at the Colorado School of Mines is captioned, "Green Hydrogen: Empowering the Future of Energy—A Closer look at hydrogen's roles in the energy transition." The article advises that hydrogen has emerged as a key player in the energy transition, identified by the International Energy Agency as a "versatile energy carrier" that has a diverse range of applications and can be deployed in a variety of sectors. But is hydrogen a moonshot? Or does it really have the potential to change the energy game? At Colorado School of Mines, teams of researchers are working on the hydrogen problem, from developing electrolyzers to separate hydrogen from other energy sources to developing and testing the ceramic materials in fuel cells and making them commercially viable and cost-effective. The research team confirms that hydrogen truly has the potential to supplement the world's energy profile in meaningful ways and help the society to successfully navigate the energy transition. The article highlights the researchers working on projects across the scope of the hydrogen approach for a deeper dive into hydrogen technology, the challenges they are facing today, how they are overcoming the challenges, and what the future looks like when powered by hydrogen. Colorado School of Mines faculty across multiple disciplines actively perform collaborative research and bring diverse perspective to the field. Close ties with industry partners enables the researchers to meet today's technological development needs. The research scope of Colorado School of Mines includes the following:

- Proton-conducting ceramic fuel cells and electrolyzers
- Solid-oxide fuel cell (SOFC) development and testing
- Fuel processing
- Modeling and simulation
- Advanced materials processing and evaluation
- Manufacturing technology development
- Systems integration

It is of particular interest that "systems integration" is included in the scope of research at the Colorado School of Mines. Systems integration happens to be a cornerstone of the approach recommended in this book through the application of the trademarked DEJI Systems Model, which is covered in Chapter 3.

The Colorado example is just one of the multitude of efforts going on around the world, either small or large. For example, the lomi.com website has a "trash-to-treasure" program that aims to address environmental issues by solving daily food waste problem, thus making ordinary families a part of the global teams of planet-conscious citizens. Lomi's participative approach suggests a role for everyone, both children and adults, in reducing food waste footprint, converting household waste to useful garden energy, using technology to divert organic waste from undesirable landfills, and so on. That is the essence of a systems approach to solving complex problems. Hendrickson et al. (2006) present a systems assessment framework for the environmental life cycle of goods and services. Their presentation aligns with the theme of this book on using systems engineering to reengineer our actions and impacts on the environment.

THE SYSTEMS ENGINEERING APPROACH

WHAT IS SYSTEMS ENGINEERING?

Systems engineering involves a recognition, appreciation, and integration of all aspects of an organization or a facility. A system is defined as a collection of inter-related elements working together in synergy to produce a composite output that is greater than the sum of the individual outputs of the components. A systems view of a process facilitates a comprehensive inclusion of all the factors involved in the process. Blanchard and Fabrycky (2006) emphasize that systems are as pervasive as the universe in which they exist. Basically, systems engineering inculcates systems science and engineering into the problem-solving domain. The systems engineering approach (Badiru, 2014) facilitates the inclusion of all pertinent elements from a technical linkage perspective. It leverages the conventional engineering problem solving methodology, as summarized here. The typical 8-step process for engineering problem solving approach consists of the following:

Step 1:   Gather data and information pertinent to the problem.
Step 2:   Develop an explicit problem statement.
Step 3:   Identify what is known and unknown.
Step 4:   Specify assumptions and circumstances.
Step 5:   Develop schematic representations and drawings of inputs and outputs.
Step 6:   Perform engineering analysis using equations and models as applicable.
Step 7:   Compose a cogent articulation of the results.
Step 8:   Perform verification, presentation, and "selling" of the result.

The steps may be tweaked, condensed, or expanded depending on the specific problem being tackled. Regardless, the steps in the framework facilitate the application of a systems-integration model. The good thing about the engineering process is that technical, social, political, economic, and managerial considerations can be factored into the solution process. The end justifies the details at hand. Systems engineering embodies the following elements:

•   Economics and evaluations of the components of the system
•   Quantitative probabilistic and statistical assessment of the system components
•   Scientific and mathematical realities governing the system
•   Hard system characteristics dealing with technical aspects
•   Soft system characteristics dealing with human elements

In the above context, a system is a set of interrelated components working together toward some common objective or purpose, which fits the goal of our climate-change response. Systems are composed of components, attributes, and relationships as summarized below:

System Components: Components are the operating parts of a system, consisting of input, process, and output. Each system component may assume a variety of values to describe a system state, as determined by some control action and one or more restrictions. The technique of state-space modeling can be applicable in this regard. Technical systems, such as electronic gadgets, consist of several intricate operating components.

System Attributes: Attributes are the properties or identifiable manifestations of the components or a system. The collection of these attributes help to characterize and describe the system. So, the system can be clearly distinguished from other systems. In some cases, similar systems may look identical on the surface, until the defining components and attributes are examined. A good example is distinguishing a male human from a female human, even though many of the defining elements are similar. Flavoring of culinary products is what can distinguish one meal item from another.

System Relationships: Relationships are the links between components and attributes. Some system elements may have one-to-one correspondence. It is an understanding of the relationships between system elements that makes systems integration possible. The following considerations set the foundation for a systems approach to operational challenges:

1. The properties and behavior of each component of the set has an effect on the properties and behavior of the set as a whole.
2. The properties and behavior of each component of the set depends on the properties and behavior of at least one other component in the set.
3. Each possible subset of components satisfies the two considerations listed above and the components cannot be divided into independent subsets.

As a complement to the above discussions, systems is embedded in the principles and practice of industrial engineering, as illustrated in the definition of **industrial engineering** below:

Industrial engineering is concerned with the design, installation, and improvement of integrated systems of people, materials, information, equipment, and energy by drawing upon specialized knowledge and skills in the mathematical, physical, and social sciences, together with the principles and methods of engineering analysis and design to specify, predict, and evaluate the results to be obtained from such systems.

The integrative process of industrial and systems engineering creates a sort of "topping out" of a project. Topping out is the term used in construction to indicate the final piece of steel being hoisted into place on a building, bridge, or other large structure. At the topping-out point, the project is not complete, but it has reached its maximum height and peripheral construction still has to continue. In our climate-change context

TABLE 1.1
Selected Tools and Techniques of Industrial and Systems Engineering

Process Management Tools

• Affinity Diagram	• DEJI Systems Model
• Benchmarking	• Flowchart
• Block Diagram	• Graph
• Brainstorming	• Interview
• Cause and Effect Diagram	• Multivoting
• Cause and Effect	• Nominal Group Technique
• Force Field Analysis	• Pareto Diagram
• Control Chart	• Q-MAP
• Customer Needs Analysis	• Survey
• Customer/Supplier Questionnaire	• Tree Diagram

Process Improvement Tools

• Action Plan	• Flowchart
• Block Diagram	• Graphs
• Brainstorming	• Histogram
• Cause and Effect Diagram	• Interview
• Cause and Effect Analysis	• Multivoting
• Control Chart	• Nominal Group Technique
• Customer Needs Analysis	• Pareto Diagram
• Customer/Supplier Modeling	• Problem Definition Checklist
• Customer/Supplier Questionnaire	• Process Map
• Data Collection Form	• Survey
• Decision Matrix	• Theory of Constraints
• DMAIC, SIPOC, PDCA, etc.	• Triple C Model

here, it means that we are instituting a systems engineering topping out, which then forms the platform for continuing the other related efforts. A systems engineering approach provides workable intersections of engineering, technology, science, mathematical modeling, ethics, socio-economic considerations, and legal implications.

The industrial and systems engineering (Badiru, 2014) approaches use a variety of tools, techniques, models, principles, and methodologies, many of which are covered in this book. Nasr (2023) highlights the importance of getting actions done urgently regarding responses to climate change. In spite of the several attempts of COPs (Conference of Parties), including COP27 (in Egypt, November 2022), many key agreements are still languishing in implementation. Industrial and systems engineering approaches can help in this regard. As an upfront summary, Table 1.1 presents selected tools and techniques of industrial and systems engineering that can be brought to bear on environmental sustainability and climate change challenges. Obviously, users can add to the list as opportunities develop and new tools emerge.

SYSTEMS OF SYSTEMS IN THE ENVIRONMENT

In the context of applying systems engineering to climate-change response, the definition of a system is not complete unless we consider its position in the hierarchical structure of the overall systems making up the environment. Every system is made up of components, and any components can be broken down into smaller chunks of components. If two hierarchical levels are involved in a given system, the lower-level hierarchy is referred to as a subsystem. In any environmental sustainability or climate-change scenario, it is important to define the system under consideration by specifying its limits, boundaries, constraints, and scope. Everything that is located outside of the system is considered as the external environment of the system. We should, however, note that no system is completely closed and isolated from its environment. Material, energy, information, and/or influence (soft or hard) must often pass back and forth through the system boundaries. The level of granularity defined for a system will determine the level of control that can be exercised on the system from external or internal sources. The term SoS (System of Systems) has emerged as an effective way of managing complex systems. For the purpose of this book, climate change consists of a multitude of systems, subsystems, and system of systems. NASA (2007) presents several examples of system of systems related to research and development in space exploration. In exploring system of systems characteristics, some of the attributes of interest include security, reliability, accessibility, safety, agility, and adaptability.

PREEMPTION IS BETTER THAN CORRECTION

Systems engineering provides an opportunity to practice the concept of "preemption is better than correction" in a way similar to the proverbial saying of prevention is better than cure. If we can preempt an adverse impact on the environment in the first place, we can contribute positively to preserving our planet. Every little damage adds up to decimating the overall environment. A case example is the hazardous train derailment in East Palestine, Ohio, USA, on February 3, 2023. The train derailment and subsequent spill of thousands of gallons of chemicals into the land and water near East Palestine led to fears among its residents that the crash may affect their health. The Norfolk Southern train had about 150 cars, of which 20 were carrying hazardous materials. The crash resulted in a fire that burned chemicals in the derailed train cars. Initial assessment was that the accident could have been prevented. Only time can tell the level and longevity of the adverse environmental impact of the train accident. Typical fallouts from such scenarios include political insinuations and accusations. Disastrous impacts on the environment can happen due to a variety of causes include the following:

- Catastrophic weather event
- Incompetence
- Mismanagement
- Sabotage
- Mischief
- Terrorism
- Deliberate act
- Negligence
- Distraction
- Natural accident

A priori systems engineering approaches can help with preemptive processes and actions to prevent disasters, or mitigate the adverse impacts.

DYNAMICS OF CLIMATE CHANGE

No matter what we believe and conjecture about climate change, the reality is that something feels different in the human experience over the preceding centuries in the history of our Earth. From a personal perspective, growing up in Lagos, Nigeria, in the 1950s, 1960s, and 1970s, I can tell the difference in the environmental temperature, as I can feel it. To put things in general perspectives, I recall a narrative analogy presented by the "Ask Marilyn" column in the *Parade Magazine* some years ago. The question addressed was about the importance of human history. The columnist's response was that the chronicle of historical events teaches us, bit by bit by bit, about civilization, human nature, human actions, and their implications and consequences. We were told to imagine two men, after a centuries-long journey in time: One was asleep all the way; the other was awake throughout the journey. The one who stayed awake witnessed and experienced every catastrophe that occurred and every success that happened. He learned where triumphs occurred and how to leverage them. He saw where and how evil developed and how to avoid it. He learned the differences between good and evil. He knew what to avoid and what to embrace. By comparison, the somnolent man has no idea what has transpired and has no basis for a comparative analysis of before and after. Which of these two men should we select as our leader? Only historical accounts can guide us toward the better decision. Do we want the man who saw everything or the man who saw nothing? Thus, the physics of climate change can embody both the physical aspects of change and the sociological aspects of our experiences and reactions.

The consensus is that human activities are responsible for climate change. For billions of years, the elements of the world had co-existed in a mutual balance, precarious as it might have been. With more and more human activities generating increasing amounts of carbon into the atmosphere, the precarious world atmospheric balance is being assaulted and disrupted. Unfortunately, the reality of life is for humans to burn fossil fuels, which generates the by-product of environmentally-adverse increases of carbon in the atmosphere. Global warming is one immediate notice of a changing climate around us. We are all experiencing more extreme and more frequent weather patterns, manifested in harsh winds, huge flooding, extreme temperatures, unusual cold spells, uncontrollable raging fires, and so on. The delicate balance of our environment is being disrupted right in front of your eyes, above our heads, and below our feet. Figure 1.1 illustrates the precarious enmeshment of previously-discernible components of the environment.

IS IT BLAH, BLAH, BLAH ALL OVER AGAIN?

As a young climate activist said (Thunberg, 2021), it seems that the world leaders are engaging in more blah, blah, blah and more blah, blah, blah, all over again, with no real actions. Sustainable world actions have been few and far between, prompting the activist to characterize world actions as just more blah, blah, blah. What do we need to do collectively to remediate the recurrence of blah, blah, blah? The approach of this book is to bring the tools and techniques of industrial and systems engineering to bear on the problems of climate change. We often tend to limit our understanding

FIGURE 1.1   A turbulent mix of land, sea, and air decimated by climate change.

and actions to the scientific frameworks of addressing climate change. However, this book suggests a broader systems-based approach encompassing humans, processes, tools, technologies, and science.

In the context of systems engineering coverage, Figure 1.2 suggests a comprehensive systems-based framework of research, policy, and education as the launch pad

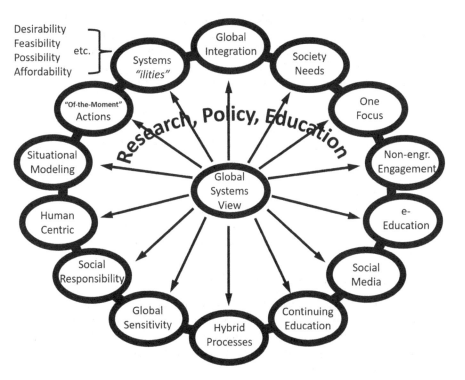

FIGURE 1.2   Wheel of systems thinking for climate change research, policy, and education.

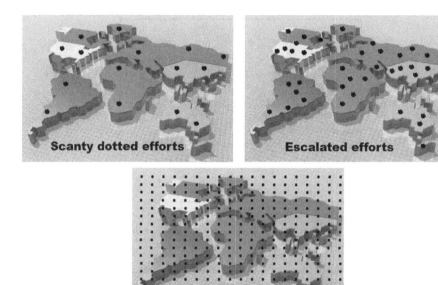

FIGURE 1.3   World systems approach to remediation of climate blah, blah, blah.

for the diverse actions needed to address climate change. A climate resolution will not happen in one fell swoop. We must continue to chip away gradually at the problem, with each action feeding into additional complementary actions. Education is one strong element of the systems engineering approach because it can foster better understanding of what is going on and what individual actions are required.

OUR CLIMATE AS A SYSTEM

A system is a collection of interrelated subsystems, whose collective output is greater than the sum of the outputs of the subsystems. In the context of the theme of this book, that output is the climate. A world system affects all of us in all corners of the world, which points to the need for systems engineering. We must address climate change from a world systems perspective. Figure 1.3 illustrates a progressive view of how climate-change responses can be made to dot and cover more and more of the world regions in a collective drive towards a sustainable solution.

A GUIDING PLATFORM FOR ACTION

In 2008, the US National Academy of Engineering released the Fourteen Grand Challenges for Engineering (NAE, 2008). Several of the topics covered by the challenges have environmental and climate implications. The list provided below paves the way for the type of broad-system considerations of several aspects affecting us, our environment, and our planet.

1. **Make solar energy economical:** Solar energy provides less than one percent of the world's total energy, but it has the potential to provide much, much more.
2. **Provide energy from fusion:** Human-engineered fusion has been demonstrated on a small scale. The challenge is to scale up the process to commercial proportions, in an efficient, economical, and environmentally benign way.
3. **Develop carbon sequestration methods:** Engineers are working on ways to capture and store excess carbon dioxide to prevent global warming.
4. **Manage the nitrogen cycle:** Engineers can help restore balance to the nitrogen cycle with better fertilization technologies and by capturing and recycling waste.
5. **Provide access to clean water:** The world's water supplies are facing new threats; affordable, advanced technologies could make a difference for millions of people around the world.
6. **Restore and improve urban infrastructure:** Good design and advanced materials can improve transportation and energy, water, and waste systems, and also create more sustainable urban environments.
7. **Advance health informatics:** Stronger health information systems not only improve everyday medical visits, but they are essential to counter pandemics and biological or chemical attacks.
8. **Engineer better medicines:** Engineers are developing new systems to use genetic information, sense small changes in the body, assess new drugs, and deliver vaccines.
9. **Reverse-engineer the brain:** The intersection of engineering and neuroscience promises great advances in health care, manufacturing, and communication.
10. **Prevent nuclear terror:** The need for technologies to prevent and respond to a nuclear attack is growing.
11. **Secure cyberspace:** It's more than preventing identity theft. Critical systems in banking, national security, and physical infrastructure may be at risk.
12. **Enhance virtual reality:** True virtual reality creates the illusion of actually being in a difference space. It can be used for training, treatment, and communication.
13. **Advance personalized learning:** Instruction can be individualized based on learning styles, speeds, and interests to make learning more reliable.
14. **Engineer the tools of scientific discovery:** In the century ahead, engineers will continue to be partners with scientists in the great quest for understanding many unanswered questions of nature.

Many of the topics are intertwined. If we make solar energy economical, we can lessen the need to use fossil fuels, thereby lessening the pressure on the environment. Ditto, if we make energy available through fusion. If we succeed in developing carbon sequestration methods, we can contribute to the remediation of global warming.

If we can manage the nitrogen cycle effectively, we can move towards restoring the balance of critical environmental ingredients. If we provide clean water to more of our population, we can impede environmentally damaging acts to obtain drinking water. If we improve urban infrastructure, we can slow down the decaying impacts of building materials that decimate the environment, with respect to transportation movements, energy consumption, waste generation, and waste disposal. If we advance health informatics, we create opportunities for cleaner living for humans, thus reducing chances for the occurrence and spread of pandemics. If we can engineer better medicines, we facilitate a healthier society that is less susceptible to adverse genetics. If we can reverse-engineer the brain, we can strengthen the intersection of human existence and the advancement of society. If we can prevent nuclear terror, we lessen the risk of environmental disaster from a nuclear war. If we can secure cyberspace, we create a more responsible society that is less subject to damaging acts on critical infrastructure. If we can enhance virtual reality, we can better simulate what the future portends for the purpose of creating preventive actions. If we can advance personalized learning, we can expand knowledge and belief in how the environment is impacted by human actions. If we can engineer dependable tools of scientific discovery, we can dig deeper into the interplay of social, economic, scientific, and technological assets affecting the environment. The efficacy of using a systems approach to influence our planet positively is that if we all work together, each one of us will have less work to do. That is, little contributions from many people translate to a large contribution by all. If we want to go far, we must go together, from a systems engineering perspective.

As NAE (2008) reminds us, throughout human history, engineering has driven the advance of civilization. From the metallurgists who ended the Stone Age to the shipbuilders who united the world's peoples through travel and trade, the past witnessed many marvels of engineering prowess. As civilization grew, it was nourished and enhanced with the help of increasingly sophisticated tools for agriculture, technologies for producing textiles, and inventions transforming human interaction and communication. Inventions such as the mechanical clock and the printing press irrevocably changed civilization (NAE, 2008). In the modern era, the Industrial Revolution brought engineering's influence to many parts of human endeavors (Badiru and Omitaomu, 2023). Machines supplemented and replaced human labor for many production activities. Improved systems for sanitation enhanced human health. The steam engine facilitated transportation and supply chains. New energy-generating techniques provided benefits for factories to increase manufacturing yield and improve product quality.

STRATEGIC PATHWAY FOR ACTION

Under the purview of NAE's 14 grand challenges, topics of urgent interest include research, policy, education, communication, cooperation, collaboration, and coordination, from a systems viewpoint. Figure 1.4 presents the comprehensive span of strategic and multifaceted actions needed to address climate change problems.

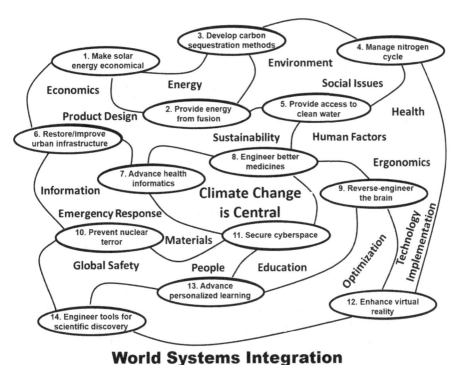

World Systems Integration

FIGURE 1.4   Climate change response mapping.

There is a potential for collaborative actions in every topic included in the possible roles of engineers and the methodology of systems engineering in addressing the requirements in NAE's grand challenges, as enumerated below.

1. **Advance Personalized Learning**
   A growing appreciation of individual preferences and aptitudes has led toward more "personalized learning," in which instruction is tailored to a student's individual needs. Given the diversity of individual preferences, and the complexity of each human brain, developing teaching methods that optimize learning will require engineering solutions of the future. The emergence of COVID-19 has made this topic even more relevant and urgent due to the need to embrace remote learning.

2. **Make Solar Energy Economical**
   As of 2008, when the list was first published, solar energy provided less than one percent of the world's total energy, but it has the potential to provide much, much more. There has been some increase in the percentage of solar usage, but there is room for more. We need to continue our systems-driven efforts in this endeavor.

3. **Enhance Virtual Reality**

   Within many specialized fields, from psychiatry to education, virtual reality is becoming a powerful new tool for training practitioners and treating patients, in addition to its growing use in various forms of entertainment. Mixed-mode simulation of production systems has been practiced by industrial engineers for decades. The same operational wherewithal can be applied to enhancing virtual reality in the context on the new world order in business, industry, academia, and government.

4. **Reverse-Engineer the Brain**

   A lot of research has been focused on creating thinking machines—computers capable of emulating human intelligence. However, reverse-engineering the brain could have multiple impacts that go far beyond artificial intelligence and will promise great advances in health care, manufacturing, and communication. The same technical principles of reverse-engineering, which business and industry already use, can be applied to human anatomy and physiological challenges.

5. **Engineer Better Medicines**

   Engineering can enable the development of new systems to use genetic information, sense small changes in the body, assess new drugs, and deliver vaccines to provide health care directly tailored to each person. The furious worldwide search for a vaccine for COVID-19 comes to mind in this regard. We need ways to engineer and develop better medicines quickly, effectively, and efficiently. Such a process is the bastion of industrial engineers, who can work with a diverse collection of scientists to expedite and secure production processes. Industrial engineering tools, such as lean production and Six Sigma are very much needed for engineering better medicines.

6. **Advance Health Informatics**

   As computers have become available for all aspects of human endeavors, there is now a consensus that a systematic approach to health informatics—the acquisition, management, and use of information in health—can greatly enhance the quality and efficiency of medical care and the response to widespread public health emergencies. The rapid emergence of new disciplines in data analytics and data science is in line with this stated area of need included in the 14 grand challenges.

7. **Restore and Improve Urban Infrastructure**

   Infrastructure is the combination of fundamental systems that support a community, region, or country. Society faces the formidable challenge of modernizing the fundamental structures that will support our civilization in centuries ahead. Industrial engineers have facilities design and urban infrastructure planning in their skills set. Thus, there is an alignment between this area of urgent need and the ready capabilities of industrial engineers.

8. **Secure Cyberspace**

   Computer systems are involved in the management of almost all areas of our lives; from electronic communications, and data systems, to controlling traffic lights to routing airplanes. It is clear that engineering needs to develop innovations for addressing a long list of cybersecurity priorities.

The increasing threat to cyberspace and the risk of unexpected hacking and authorized data incursion have made it imperative to expedite research and development in cyberspace security.

9. **Provide Access to Clean Water**
   The world's water supplies are facing new threats; affordable, advanced technologies could make a difference for millions of people around the world. Efficient management of resources is an area of interest and expertise for industrial engineers. Water is a resource that we often take for granted. With a systems-oriented view of all resources, industrial engineers can be instrumental in the global efforts to provide access to clean water.

10. **Provide Energy from Fusion**
    Human-engineered fusion has been demonstrated on a small scale. The challenge is to scale up the process to commercial proportions, in an efficient, economical, and environmentally benign way. This is a highly technical area of pursuit. But it is still an area that can benefit from better managerial policies and procedures, from the perspectives of industrial and systems engineering. No matter how technically astute a system might be, it will still need good policies and collaborative governance. This is something that industrial engineers can contribute to in the pursuit of providing energy from alternate sources.

11. **Prevent Nuclear Terror**
    The need for technologies to prevent and respond to a nuclear attack is growing. The threat for a nuclear attack is always human-driven. The more we can understand how the other side thinks and operates, the better we can a handle on how to prevent nuclear terror. The human-focused practice of industrial engineering may have something to offer with respect to understanding human behavior, reactions, and tendencies. Both preparation and prevention are essential in the endeavor to better leverage nuclear capabilities while limiting the chances for terror.

12. **Manage the Nitrogen Cycle**
    Engineers can help restore balance to the nitrogen cycle with better fertilization technologies and by capturing and recycling waste. Managing the nitrogen cycle requires management, from the standpoint of technical requirements and human sensitivities. The reuse, recover, and recycle programs that industrial engineers often participate in can find a place in this topic.

13. **Develop Carbon Sequestration Methods**
    Engineers are working on ways to capture and store excess carbon dioxide to prevent global warming. Industrial engineers already work within domain of designing efficient storage systems, whether for industrial physical assets or non-physical resources. It is all about the strategic placement of things, while considering the systems influences coming from other things.

14. **Engineer the Tools of Scientific Discovery**
    In the century ahead, engineers will continue to be partners with scientists in the great quest for understanding many unanswered questions of nature. The toolbox of industrial engineers contains what it takes to advance and leverage the tools and processes for scientific discovery.

Our society will be tackling these grand challenges for the foreseeable decades; and project management is one avenue through which we can ensure that the desired products, services, and results can be achieved. With the positive outcomes of these projects achieved, we can improve the quality of life for everyone and our entire world can benefit positively. In the context of tackling the grand challenges as system-based projects, some of the critical issues to address are

- Strategic implementation plans
- Strategic communication
- Knowledge management
- Evolution of virtual operating environment
- Structural analysis of projects
- Analysis of integrative functional areas
- Project concept mapping
- Prudent application of technology
- Scientific control
- Engineering research and development

Much research remains to done. Research must be followed by sustainable development. NAS (2021) presents a summary of the climate-change research needs and opportunities for 2022 to 2031. A systems view of energy security is a focused item on the research agenda highlighted in the consensus report. The other key items addressed in the report are summarized in Table 1.2.

THE ESSENCE OF SYSTEMS INTEGRATION

We must integrate all the elements of a project on the basis of alignment of functional goals. Systems overlap for integration purposes can conceptually be represented as

TABLE 1.2
Climate-Change Research Needs and Opportunities for 2022–2031

Areas of Focus	Systems Studies
Systems	**Natural Systems:** physical climate system and ecosystems; dynamics governed by biophysical and/or physical processes **Human Systems:** systems managed by people to meet specific needs of society; dynamics governed by human actions **Coupled Human–Natural Systems:** systems with interconnected, interdependent, and complex interactions among human systems, the physical climate system and ecosystems
Risks	**Risks:** the potential for adverse consequences for human or natural systems from exposure to hazards **Integrated Risks:** risks to coupled human-natural systems
Risk Management Approaches	**Management Approaches:** strategies to minimize risks to systems; include mitigation, adaptation, and strategies that combine them **Integrated Risk Management Approaches:** address risks to human and natural systems, as well as synergies and tradeoffs, to increase security of the integrated system

projection integrals by considering areas bounded by the common elements of subsystems. Multidisciplinary education is essential in grasping the integrated concepts and principles that exist among the elements of the 14 grand challenges for engineering.

In response to a need to actualize NAE's 14 grand challenges in the context of engineering education, Badiru (2013, 2015) presents Fifteen Grand Strategies for global engineering education on climate change. The elements of his strategies are summarized below:

1. Systems view of the world in educational delivery modes and methods in order to leverage unique learning opportunities around the world.
2. Pursuit of integration and symbiosis of global academic programs. Through global educational system integration, we can move toward a mutual-assured advancement of engineering education. Think global, but educate locally to fit domestic needs. Language diversity, for example, can expand thought and understanding to facilitate global communication, cooperation, and coordination.
3. Linking engineering education to the present and future needs of society rather than just a means to better employment.
4. Commitment to embrace all engineering disciplines in a collaborative one-focus alliance toward addressing societal challenges.
5. Engagement of non-engineering disciplines, such as management and the humanities, in addressing high-value societal problems collectively. There are now medical humanities programs. Consider engineering humanities programs to put a human face to engineering solutions.
6. Adoption and adaptation of e-education to facilitate blended learning modes, flexibility of learning, and diversity of thought in a fast-paced society. Of interest in this regard is the evolution of measurement scales for pedagogy and andragogy.
7. Leveraging of social media tools and techniques to facilitate serious and rigorous transmission of knowledge.
8. Extension of formal engineering education to encompass continuing engineering education and sustainability of learning.
9. Creation of hybrid method of teaching what is researched and researching what is taught.
10. Inculcation of global sensitivity into engineering education programs.
11. Inclusion of social responsibility in engineering education, research, and practice.
12. Making engineering solutions more human-centric solutions. Use engineering to solve real human problems. Keep engineering education relevant to the needs of society.
13. Teaching of representational modeling in engineering education. Modeling can provide historical connectivity to recognize the present as an output of the past and a pathway for the future.
14. Teaching of "Of-the-Moment Creativity" to spur innovation for the current, prevailing, and attendant problem.
15. Introduction of engineering solution "ilities" covering feasibility, sustainability, viability, desirability of engineering solution approaches.

TABLE 1.3
Taxonomical Analysis of Research, Education, and Practice Alignment in NAE's Grand Challenges

NAE's 14 Grand Challenges	Research Topics	Education Topics	Practice Topics
1. Make solar energy economical	Strategic Investments	Engineering Economics	Portfolio Management
2. Provide energy from fusion	Design Safety	Energy Management	Energy Strategies
3. Develop carbon sequestration methods	Natural Science Analytics	Storage Systems Design	Global Systems Interfaces
4. Manage the nitrogen cycle	System Planning	Systems Optimization	Environmental Science
5. Provide access to clean water	Environmental Science	Water Resource Engineering	Water Management & Remediation
6. Restore and improve urban infrastructure	Resilience Engineering	Construction Management	Infrastructure Design
7. Advance health informatics	Health Systems Engineering	Computer Science & Bio informatics	Health Informatics
8. Engineer better medicines	Personalized Pharmaceuticals	Biomedical Engineering	Health Systems Engineering in Medicine
9. Reverse-engineer the brain	Human Factors	Cognitive Psychology	Behavioral Science
10. Prevent nuclear terror	Deterrent Strategies	Emergency Response	Social Responsibility and Negotiation
11. Secure cyberspace	Information Resources Research	Enterprise Design and Management	Information Security
12. Enhance virtual reality	Software Engineering	Software Design and Programming	Game Design
13. Advance personalized learning	Active Learner Systems	Blended Hybrid Learning	Distance Learning Systems
14. Engineer the tools of scientific discovery	Product Design	Scientific Inquiry	Ergonomics

Badiru (2014, 2019a, 2019b, 2020, 2023) presents various tools and techniques of industrial and systems engineering for accomplishing the strategic elements outlined above. Some of these are addressed fully in subsequent chapters of this book. Education is the avenue through which the goals and objectives of the 14 grand challenges can be realized. It could be education in terms of preparing the future engineers or education in the form of raising awareness of members of the society. Table 1.3 presents a taxonomy of education implications of the 14 grand challenges.

Energy, in its various forms, appears to be a common theme in many of the 14 grand challenges. In addition to teaching the technical and analytical topics related to energy, energy-requirement analysis must consider the social and cultural aspects involving the three primary focus areas listed below.

- **Energy generation:** Fossil, fission, fusion, renewable, non-renewable, safety, security, energy independence
- **Energy distribution:** Transmission technology, hydrogen, distributed energy sources, market
- **Energy consumption:** Transportation, storage, product requirements, conservation, recovery, recycling

These focus areas give us, as individuals and groups, multiple opportunities to reengineer our actions in ways to positively influence our planet, which is the premise of this book.

As a strategy, climate-centric education must be delivered within the proper systems context. Below are some selected educational headlines: Focus on Institutional Culture Drives High Rate of Faculty Technology Adoption at Genesee CC

- 5 Universities Launch Engineering Research Centers with $92.5 Million in NSF Funding
- Louisiana Researchers Connect with 600 Megabit Ethernet Deployment

When you have the right tools, things come into focus.

Education misapplied is education missed. A systems model that can help ensure effective design, evaluation, justification, and integration is the DEJI model (Badiru, 2019b, 2023). The model has been applied to product quality management as well as aerospace product development (Badiru, 2012) to effect the cycle of

Design–evaluation–justification–integration in the operating environment. Education is the ultimate operating environment where several subsystems must interact to bring about a successful end product. Some unique education environments require specialized views and practices. For example, in military education systems, a hybrid credit assessment process leverages military service for advancing STEM career development. In general, every aspect of engineering education, from curriculum development to educational outcome assessment, can benefit from a structured application of a systems model. Using climate-focused curriculum development as an illustrative example, consider the following stages:

Design of Curriculum: An educational curriculum must be designed to fit the current technological tools and the prevailing needs in the job market.

Evaluation of Curriculum: The curriculum must be evaluated periodically to ensure that it continues to meet the identified needs of the education process.

Justification of Curriculum: There should not be a de facto curriculum. Every curriculum should be evaluated and justified to ensure a continuity alignment with educational objectives.

Integration of Curriculum: Each educational outcome must be integrated to the prevailing needs of the society. What is desirable and practical in one system may not be realistic in other systems. To minimize disconnect, the local environment of education must be considered when designing, evaluating, and justifying new or existing curricula. For example, the social environment, economic scenario, cultural needs, and political requirements are all essential within the systems view of global engineering education. Among the critical issues to be brought to the forefront as a result of applying a structural model to engineering education are the following:

- Focus on learner-centered curriculum design
- Pursuit of transformational engineering education that is aligned with the prevailing times
- Recognition of active and passive learning modes
- Perspective of a holistic view of engineering education input and output processes
- Essentials of teaching with the latest technology
- Appreciation of the cultural, social, political, and economic limitations affecting engineering education in some nations
- Governmental competence in initiating and/or supporting engineering education programs
- Appreciation of hybrid education resources, including distance learning and online resources
- Emergence of new engineering education standards as a result of structural design, evaluation, justification, and integration of engineering education processes

- Encouragement of engineering education outcomes through diversity of backgrounds and views
- Facilitation of self-motivational and self-regulated student learning

TAKING PERSONAL ACCOUNTABILITY FOR ACTION

Demonstrating personal accountability for actions related to climate change is essential for a global success in influencing our planet positively. Personal commitment to sustainability is one ready-made approach to making a contribution. Sustainability is not just for the environment. Although environmental concern is what immediately comes to mind whenever the word "sustainability" is mentioned, there are many languages (i.e., modes) of sustainability, depending on whatever perspective is under consideration. The context determines the interpretation. Each point of reference determines how we, as individuals or groups, respond to the need for sustainability. Pursuits of green building, green engineering, clean water, climate research, energy conservation, eco-manufacturing, clean product design, lean production, and so on remind us of the foundational importance of sustainability in all we do.

Commitment to sustainability is in vogue these days, be it in the corporate world or personal pursuits. But, what exactly is sustainability? Definitions of the word contain verbs, nouns, and adjectives such as "green," "clean," "maintain," "retain," "stability," "ecological balance," "natural resources," and "environment." The definition of sustainability implies the ability to sustain (and maintain) a process or object at a desirable level of utility. The concept of sustainability applies to all aspects of functional and operational requirements, embracing both technical and managerial needs. Sustainability requires methodological, scientific, and analytical rigor to make it effective for managing human activities and resources.

In the above context, sustainability is nothing more than prudent resource utilization. The profession of industrial engineering is uniquely positioned to facilitate sustainability, especially as it relates to the environment, technical resources, management processes, human interfaces, product development, and facility utilization. Industrial engineers have creative and simple solutions to complex problems. Sustainability is a complex undertaking that warrants the attention and involvement of industrial engineers. A good example of the practice of sustainability is how a marathon runner strategically expends stored energy to cover a long-distance race. Burning up energy too soon means that the marathon race will not be completed. Erratic expenditure of energy would prevent the body from reaching its peak performance during the race. Steady-state execution is a foundation for achieving sustainability in all undertakings where the decline of an asset is a concern. An example is an analysis of how much water or energy it takes to raise cattle for human consumption. Do we ever wonder about this as we delve into our favorite steaks? Probably not. Yet, this is, indeed, an issue of sustainability. For this reason, we present the following succinct definition of the role of industrial engineers in sustainability:

Industrial engineers make systems function better together with less waste, better quality, fewer resources, and on target with goals, objectives, and requirements.

EXERCISING RESOURCE CONSCIOUSNESS

This section is in alignment with the theme of dedicating this book to "Aina," the Hawaiian word for "land" and its value to human existence. The often-heard debate about what constitutes sustainability can be alleviated if we adopt the context of "resource consciousness," which, in simple terms, conveys the pursuit of conservation in managing our resources. All the resources that support our objectives and goals are amenable to sustainability efforts. For example, the expansion of a manufacturing plant should consider sustainability, not only in terms of increased energy consumption, but also in terms of market sustainability, intellectual property sustainability, manpower sustainability, product sustainability, and so on. The limited resource may be spread too thin to cover the increased requirements for a larger production facility. Even a local community center should consider sustainability when contemplating expansion projects just as the local government should consider tax base sustainability when embarking on new programs. The mortgage practices that led to the housing industry bust in the US were due to financial expectations that were not sustainable. If we put this in the context of energy consumption, it is seen that buying a bigger house implies a higher level of energy consumption, which ultimately defeats the goal of environmental sustainability. Similarly, a sports league that chooses to expand haphazardly will eventually face a non-sustainability dilemma. Every decision ties back to the conservation of some resource (whether a natural resource or a manufactured resource), which links directly to the conventional understanding of sustainability.

PART AND PARCEL OF SUSTAINABILITY

There are several moving parts in sustainability. Only a systems view can ensure that all components are factored into the overall pursuit of sustainability. A systems view of the world allows an integrated design, analysis, and execution of sustainability projects. It would not work to have one segment of the world embarking on sustainability efforts while another segment embraces practices that impede overall achievement of sustainability. In the context of production for the global market, whether a process is repeatable or not, in a statistical sense, is an issue of sustainability. A systems-based framework allows us to plan for prudent utilization of scarce resources across all operations. Some specific areas of sustainability include the following:

- Environmental sustainability
- Operational sustainability
- Energy sustainability
- Health and welfare sustainability
- Safety and security sustainability
- Market sustainability
- Financial sustainability
- Economic sustainability
- Health sustainability
- Family sustainability
- Social sustainability

The long list of possible areas means that sustainability goes beyond environmental concerns. Every human endeavor should be planned and managed with a view toward sustainability.

SUSTAINING VALUE THROUGH SUSTAINABILITY

Sustainability imparts value on any organizational process and product. Even though the initial investment and commitment to sustainability might appear discouraging, it is a fact that sustainability can reduce long-term cost, increase productivity, and promote achievement of global standards. Sample questions for value sustainability are provided below:

- What is the organizational mission in relation to the desired value stream?
- Are personnel aware of where value resides in the organization?
- Will value assignment be on team, individual, or organizational basis?
- Is the work process stable enough to support the acquisition of value?
- Can value be sustained?

LEVERAGING PERSONAL HIERARCHY OF NEEDS

Sustainability is often tied to the basic needs of humans. The psychology theory of "Hierarchy of Needs" postulated by Abraham Maslow in his 1943 paper, "A Theory of Human Motivation," still governs how we respond along the dimensions of sustainability, particularly where group dynamics and organizational needs are involved. An environmentally-induced disparity in the hierarchy of needs implies that we may not be able to fulfill personal and organizational responsibilities along the spectrum of sustainability. In a diverse workforce, the specific levels and structure of needs may be altered from the typical steps suggested by Maslow's Hierarchy of Needs. This calls for evaluating the needs from a multidimensional perspective. For example, a 3-D view of the hierarchy of needs can be used to coordinate personal needs with organizational needs with the objective of facilitating sustainability. People's hierarchy of basic needs will often dictate how they respond to calls for sustainability initiatives. Maslow's hierarchy of needs consists of five stages:

1. **Physiological Needs:** These are the needs for the basic necessities of life, such as food, water, housing, and clothing (i.e., survival needs). This is the level where access to money is most critical. *Sustainability applies here.*
2. **Safety Needs:** These are the needs for security, stability, and freedom from physical harm (i.e., desire for a safe environment). *Sustainability applies here.*
3. **Social Needs:** These are the needs for social approval, friends, love, affection, and association (i.e., desire to belong). For example, social belonging may bring about better economic outlook that may enable each individual to be in a better position to meet his or her social needs. *Sustainability applies here.*
4. **Esteem Needs:** These are the needs for accomplishment, respect, recognition, attention, and appreciation (i.e., desire to be known). *Sustainability applies here.*

5. **Self-Actualization Needs:** These are the needs for self-fulfillment and self-improvement (i.e., desire to arrive). This represents the stage of opportunity to grow professionally and be in a position to selflessly help others. *Sustainability applies here.*

Ultimately, the need for and commitment to sustainability boil down to each person's perception based on his or her location on the hierarchy of needs and level of awareness of sustainability. How do we explain to a hungry, poor family in an economically depressed part of the world the need to conserve forestry? Or, how do we dissuade an old-fashioned professor from the practice of making volumes of hardcopy handouts instead of using electronic distribution? Cutting down on printed materials is an issue of advancing sustainability. In each wasteful eye, "the *need* erroneously justifies the *means*" (author's own variation of the common phrase). This runs counter to the principle of sustainability. We can expand the hierarchy of needs to generate a multidimensional view that incorporates organizational hierarchy of needs. The location of each organization along its hierarchy of needs will determine how the organization perceives and embraces sustainability programs. Likewise, the hierarchy position of each individual will determine how he or she practices commitment to sustainability.

In an economically underserved culture, most workers will be at the basic level of physiological needs, and there may be constraints on moving from one level to the next higher level. This fact has an implication on how human interfaces impinge upon sustainability practices. In terms of organizational hierarchy of needs, the levels are characterized as follows:

Level 1 of Organizational Needs: This is the organizational need for basic essentials of economic vitality to support the provision of value for stockholders and employees. Can the organization fund projects from cash reserves? *Sustainability applies here.*
Level 2 of Organizational Needs: This is a need for organizational defense. Can the organization feel safe from external attack? Can the organization protect itself cyber-attacks or brutal take-over attempts? *Sustainability applies here.*
Level 3 of Organizational Needs: This is the need for an organization to belong to some market alliances. Can the organization be invited to join trade groups? Does the organization have a presence on some world stage? *Sustainability applies here.*
Level 4 of Organizational Needs: This is the level of having market respect and credibility. Is the organization esteemed in some aspect of market, economic, or technology movement? What positive thing is the organization known for? *Sustainability applies here.*
Level 5 of Organizational Needs: This is the level of being classified as a "Power" in the industry of reference. Does the nation have a recognized niche in the market? *Sustainability applies here.*

Obviously, where the organization stands in its hierarchy of sustainability goals will determine how it influences its employees (as individuals) to embrace, support, and practice sustainability. How each individual responds to organizational requirements depends on that individual's own level in the hierarchy of needs. We must all recognize the factors that influence sustainability in our strategic planning programs. In order for an organization to succeed, sustainability must be expressed explicitly as a goal across organizational functions.

BADIRU'S SUSTAINABILITY MATRIX

The coupling of technical assets and managerial tools are essential for realizing sustainability. This section presents an example of the *sustainability matrix* introduced by Badiru (2010). The matrix is a simple tool for organizing the relevant factors associated with sustainability. It overlays sustainability awareness factors, technical assets, and managerial tools. The sample elements illustrate the nature and span of factors associated with sustainability projects. Each organization must assess its own environment and include relevant factors and issues within the context of prevailing sustainability programs. Without a rigorous analytical framework, sustainability will just be in talks rather than deeds. One viable strategy is to build collaborative STEM (Science, Technology, Engineering, and Mathematics) alliances for sustainability projects. The analytical framework of systems engineering provides a tool for this purpose from an interdisciplinary perspective. With this, environmental systems, industrial systems, and societal systems can be sustainably tied together to provide win–win benefits for all. An effective collaborative structure would include researchers and practitioners from a wide variety of disciplines (civil and environmental engineering, industrial engineering, mechanical engineering, public health, business, etc.).

Project sustainability is as much a need as the traditional components of project management spanning planning organizing, scheduling, and control. Proactive pursuit of the best practices of sustainability can pave the way for project success on a global scale. In addition to people, technology, and process issues, there are project implementation issues. In terms of performance, if we need a better policy, we can develop it. If we need technological advancement, we have capabilities to achieve it. The items that are often beyond reach relate to project life cycle management issues. Project sustainability implies that sustainability exists in all factors related to the project. Thus, we should always focus on project sustainability.

The vertical dimension of the matrix consists of the following *technical factors*:

- Physical infrastructure
- Work design
- Analytical modeling
- Scientific limitation
- Technology constraints

The horizontal dimension of the matrix consists of the following *managerial environmental factors*:

- Organizational behavior
- Personnel culture

- Resource base
- Market influence
- Share capital

The cells within the matrix consist of a variety of attributes, factors, and indicators, including the following: Communication modes, cooperation incentives, coordination techniques, building performance, energy economics, technical acquisitions, work measurement, project design, financial implications, project control, resource combinations, qualitative risk, engineering analysis, value assessment, forecast models, fuel efficiency, technical workforce, contingency planning, contract administration, green purchases, energy conservation, training programs, quantitative risk, public acceptance, and technology risks.

Think "sustainability" in all you do and you are bound to reap the rewards of better resource utilization, operational efficiency, and process effectiveness. Both management and technical issues must be considered in the pursuit of sustainability. People issues must be placed at the nexus of all the considerations of sustainability. Otherwise, sustainability itself cannot be sustained. Many organizations are adept at implementing Rapid Improvement Events (RIE). This chapter recommends a move from mere RIE to Sustainable Improvement Events (SIE).

SOCIAL CHANGE FOR SUSTAINABILITY

Change is the root of advancement. Sustainability requires change. Our society must be prepared for change in order to achieve sustainability. Efforts that support sustainability must be instituted into every aspect of everything that the society does. If society is better prepared for change, then positive changes can be achieved. The "pain but no gain" aspects of sustainability can be avoided if proper preparations have been made for societal changes. Sustainability requires an increasingly larger domestic market to preserve precious limited natural resources. The social systems that make up such markets must be carefully coordinated. The socio-economic impact on sustainability cannot be overlooked.

Social changes are necessary to support sustainability efforts. Social discipline and dedication must be instilled in the society to make sustainability changes possible. The roles of the members of a society in terms of being responsible consumers and producers of consumer products must be outlined. People must be convinced of the importance of the contribution of each individual whether that individual is acting as a consumer or as a producer. Consumers have become so choosy that they no longer will simply accept whatever is offered in the marketplace. In cases where social dictum directs consumers to behave in ways not conducive to sustainability, gradual changes must be facilitated. If necessary, an acquired taste must be developed to like and accept the products of local industry. To facilitate consumer acceptance, the quality of industrial products must be improved to competitive standards. In the past, consumers were expected to make do with the inherent characteristics of products regardless of potential quality and functional limitations. This has changed drastically in recent years. For a product to satisfy the sophisticated taste of the modern consumer, it must exhibit a high level of quality and responsiveness to the needs of the consumer with respect to global expectations. Only high-quality products and services can survive the prevailing market competition and, thus, fuel the enthusiasm for further sustainability efforts. Some of the approaches for preparing a society for sustainability changes are listed below:

- Make changes in small increments
- Highlight the benefits of sustainability development
- Keep citizens informed of the impending changes
- Get citizen groups involved in the decision process
- Promote sustainability change as a transition to a better society
- Allay the fears about potential loss of jobs due to new sustainability programs
- Emphasize the job opportunities to be created from sustainability investments

Addressing the above issues means using a systems view to tackle various challenges of executing sustainability. As has been discussed in the preceding sections, the concept of sustainability is a complex one. However, with a systems approach, it is possible to delineate some of its most basic and general characteristics. For our sustainability purposes, a system is simply defined as a set of interrelated elements (or subsystems). The elements can be molecules, organisms, machines, machine components, social groups, or even intangible abstract concepts. The relations, interlinks, or "couplings" between the elements may also have very different manifestations (e.g., economic transactions, flows of energy, exchange of materials, causal linkages, control pathways, etc.). All physical systems are *open* in the sense that they have exchanges of energy, matter and information with their environment that are significant for their functioning. Therefore, what the system "does", in its behavior, depends not only on the system itself, but also on the factors, elements or variables coming from the environment of the system. The environment impacts "inputs" onto the system while the system impacts "outputs" onto the environment. This, in essence, is the systems view of sustainability.

CONCLUSIONS

Engineering education is the foundation for solving complex problems now and in the future. In order to educate and inspire the present and future generations of engineering students to tackle the pressing challenges of the world, a global systems perspective must be pursued. The 14 grand challenges for engineering released by the National Academy of Engineering in 2008 highlight the pressing needs. But until we can come up with practical and viable models and templates, the challenges will remain only in terms of ideals rather than implementation ideas. In this book's approach, we suggest Badiru's 15 grand challenges for global engineering education, tailored to climate change, as one practical pathway for addressing NAE's 14 challenges for engineering. This opening chapter also suggests the DEJI Systems Model for a structural approach to system design, evaluation, justification, and integration. The methodology presented in this chapter and the chapters that follow is particularly appropriate for adaptation for global environmental response programs.

REFERENCES

Badiru, Adedeji B. (2010), "The Many Languages of Sustainability," *Industrial Engineer*, Vol. 2, No. 2, pp. 31–34.

Badiru, Adedeji B. (2012), "Application of the DEJI Model for Aerospace Product Integration," *Journal of Aviation and Aerospace Perspectives (JAAP)*, Vol. 2, No. 2, pp. 20–34.

Badiru, Adedeji B. (2013), "NAE's 14 Grand Challenges and 2020 Skills for Engineers," *Distinguished Keynote Address, World Congress on Engineering Education (WCEE 2013)*, Doha, Qatar, January 7–9, 2013.

Badiru, Adedeji B. editor. (2014), *Handbook of Industrial & Systems Engineering*, Second Edition, Taylor & Francis Group/CRC Press, Boca Raton, FL.

Badiru, Adedeji B. (2015), "A Systems Model for Global Engineering Education: The 15 Grand Challenges," *Engineering Education Letters*, Vol. 1, No. 1, pp. 1–14. Open Access Link: 2015:3 https://doi.org/10.5339/eel.2015.3.

Badiru, Adedeji B. (2016), "A Global Systems View of Climate Change: Research, Policy, and Education," *Presentation at the London Climate Change Conference*, Westminster, London, UK, October 25, 2016.

Badiru, Adedeji B. (2019a), *Project Management: Systems, Principles, and Applications*, Second Edition, Taylor & Francis Group/CRC Press, Boca Raton, FL.

Badiru, Adedeji B. (2019b), *Systems Engineering Models: Theory, Methods, and Applications*, Taylor & Francis Group/CRC Press, Boca Raton, FL.

Badiru, Adedeji B. (2020), *Innovation: A Systems Approach*, Taylor & Francis Group/CRC Press, Boca Raton, FL.

Badiru, Adedeji B. (2023), *Systems Engineering Using DEJI Systems Model: Design, Evaluation, Justification, and Integration with Case Studies and Applications*, Taylor & Francis Group/CRC Press, Boca Raton, FL.

Badiru, Adedeji B. and Olufemi A. Omitaomu (2023), *Systems 4.0: Systems Foundation for Industry 4.0*, Taylor & Francis Group/CRC Press, Boca Raton, FL.

Badiru, Adedeji B. and Samuel O. Osisanya (2013), *Project Management for the Oil & Gas Industry*, Taylor & Francis Group/CRC Press, Boca Raton, FL.

Badiru, Adedeji B. and Tina Agustiady (2021), *Sustainability: A Systems Engineering Approach to the Global Grand Challenge*, Taylor & Francis Group/CRC Press, Boca Raton, FL.

Blanchard, Benjamin S. and Wolter J. Fabrycky (2006), *Systems Engineering and Analysis*, Prentice-Hall, Upper Saddle River, NJ.

Gamble, Simon (2014), "Taking the Lead," *PM Network*, Vol. 28, No. 10, October, pp. 36–37.

Hendrickson, Chris T., Lester B. Lave, and H. Scott Matthews (2006), *Environmental Life Cycle Assessment of Goods and Services: An Input-Output Approach*, Routledge, London, UK.

Jones, Tegan (2014), "The Energy Revolution," *PM Network*, Vol. 28, No. 10, October, pp. 30–35.

Lathan, Corinna (2022), *Inventing the Future: Stories from a Techno-Optimist*, Lioncrest Publishing, Carson City, Nevada.

Mines Staff (2023), "Green Hydrogen: Empowering the Future of Energy – A Closer Look at Hydrogen's Roles in the Energy Transition," Colorado School of Mines Research Report, Golden, Colorado, pp. 18–25.

NAE (2008), *NAE Grand Challenges for Engineering*, US National Academy of Engineering, National Academies, The National Academy Press, Washington, DC.

NAS (2021), *Global Change Research Needs and Opportunities for 2022-2031: Consensus Study Report*, US National Academy of Science, National Academies, The National Academy Press, Washington, DC.

NAS (2018), *Review of the Draft Fourth National Climate Assessment: Consensus Study Report*, US National Academy of Science, National Academies, The National Academy Press, Washington, DC.

NAS and TRS (2020), US National Academy of Sciences and the British Royal Society, Climate Change Evidence and Causes: Update 2020 – An overview from the Royal Society and the US National Academy of Sciences, US National Academy of Science, National Academies, The National Academy Press, Washington, DC.

NASA (2007), *NASA Systems Engineering Handbook*, National Aeronautical and Space Administration, NASA Headquarters, Washington, DC, December.

Nasr, Nabil (2023), "Another Step Toward Climate Action," *ISE Magazine*, Vol. 55, No. 2, p. 24.

NRC (2010), *Verifying Greenhouse Gas Emissions: Methods to Support International Climate Agreements*, National Research Council, National Academies, The National Academy Press, Washington, DC.

Sambo, Abubakar S. (2023), "The Increasing Global Uptake of Renewable Energy and the Way Forward for Nigeria," *Presentation of the Renewable Energy Committee of the Nigerian Academy of Engineering*, Abuja, Nigeria, February, 21, 2023.

Thunberg, Greta (2021), "Blah, Blah, Blah: Greta Thunberg Lambasts World Leaders Over Climate Crisis," https://www.theguardian.com/environment/2021/sep/28/blah-greta-thunberg-leaders-climate-crisis-co2-emissions (accessed February 20, 2023).

2 Empirical Facts and Figures of Our Environment

PERMISSION NOTE

This chapter is an adaptation, not a direct reprint, of parts of IRENA (2023), with permission noted as follows: Unless otherwise stated, material in this publication may be freely used, shared, copied, reproduced, printed and/or stored, provided that appropriate acknowledgement is given of IRENA as the source and copyright holder.

Citation: IRENA and CPI (2023), **Global landscape of renewable energy finance**, 2023, International Renewable Energy Agency, Abu Dhabi. ISBN: 978-92-9260-523-0

ENERGY TRANSITION WITH RENEWABLES

Too Little Water of Too Much Water; It's all in the Climate System.

– Adedeji Badiru

The theme of this book is the application of systems engineering to reengineer our collective and individual actions to influence our planet positively. Energy responsiveness and responsibilities are the most obvious, direct, and ready avenues for influencing our environment, climate, and our planet. Investments in this collective approach is essential for a global success, from a systems engineering perspective. Investments, in this regard, are multifaceted and can take different forms that are combinations and permutations of the following actions:

- Direct monetary expenditure
- Research studies to better understand the problem
- Personal investment in energy-efficient assets
- Community commitment to energy-conscious programs
- Government programs to fund or support energy initiatives
- Political coalition to get everyone (or most people) on the same page
- International grants to developing countries to commit themselves further
- Global programs to enjoin developed countries to team up for a global solution
- Parity and equity initiatives to encourage each player to participate fully

- Neighborhood programs to self-police environment-affecting practices
- Academic curricula to address educational needs for sustainable solutions
- Industry tax incentives to encourage climate-focused programs
- Industrial production practices focused on energy-efficient operations
- Sweeping collaborative alliances across the spectrum of the society

All nations, all governments, all groups, and all individuals are charged with paying urgent attention to the declining state of our world. The view of this author is that a systems engineering framework offers a viable approach to tackling this global challenge. There are many scattered efforts, many of them effective and successful in their own respective rights. While all efforts should be recognized and hailed, the gaps that exist should be identified and addressed. There are many multidimensional efforts addressed diverse aspects of responding to climate change and the impacts on our planet and our very existence on Earth. Humans affect our environment and the planet in many ways, but activities related to energy use is a major noticeable part of the overall impact. The processes of seeking, exploring, and harvesting energy for our growing consumption needs pose the biggest risk to the environment. For this reason, there are growing calls for us to seek alternate energy sources. This has led to calls for renewable energy sources. This is why this chapter focuses specifically on the need to pay more direct and sustainable attention to renewable energy sources. It is through energy management activities that individuals have direct opportunities and personal control to make contributions to the issues of climate change. This is the basis of having "reengineering our actions" in the title of this book.

Renewables are at the heart of the global energy transition, a transition that promises to put the world on a climate-safe pathway while ensuring universal access to sustainable, reliable, and affordable energy. The urgent need for accelerated renewable energy investments is further underscored by the widening effects of climate change around the world, growing food shortages and the looming energy crisis. An energy sector based on renewables can offer improved energy security and independence, price stability and reductions in greenhouse gas emissions, all of which are required to achieve climate and sustainable development goals. To that end, significant capital must be shifted from fossil fuels to renewables at a faster pace. Investments in renewables must more than triple from their current level. The major renewable energy sources are:

1. Solar
2. Wind
3. Hydro
4. Biomass and
5. Geothermal

Renewable energy sources, whose supply never runs out, are convertible to electricity, on-grid or off-grid, and to heat through the following:

- Photovoltaic (PV)
- Concentrated Solar Power (CSP)

A 2022 report on the Global Landscape of Renewable Energy documents that global investment in energy transition technologies reached US$1.3 trillion in 2022. This includes energy efficiency programs. The 2022 investment is a new record and represents a 19% increase above the 2021 investment and a 50% increase from the 2019 level, before the COVID-19 pandemic.

The joint report by the International Renewable Energy Agency (IRENA) and Climate Policy Initiative (CPI)—launched on the sidelines of the Spanish International Conference on Renewable Energy in Madrid—also finds that, although global investment in renewable energy reached a record high of US$0.5 trillion in 2022, this still represents less than 40% of the average investment needed each year between 2021 and 2030, according to IRENA's 1.5°C global temperature increase scenario. Investments are also not on track to achieve the goals set by the 2030 Agenda for Sustainable Development.

Since decentralized solutions are vital in plugging the access gap to reach universal energy access to improve livelihoods and welfare under the 2030 Agenda, efforts must be made to scale up investments in the off-grid renewables sector. Despite reaching record-high annual investments exceeding US$0.5 billion in 2021, investment in off-grid renewable solutions falls far short of the US$2.3 billion needed annually in the sector between 2021 and 2030.

For us to be successful for the long term, global investment in renewables must double to meet climate and development goals. This report provides recommendations to scale up public funds and channel them more towards developing economies.

Furthermore, investments have become concentrated in specific technologies and uses. In 2020, solar photovoltaic alone attracted 43% of the total investment in renewables, followed by onshore and offshore wind at 35% and 12% shares, respectively. Based on preliminary figures, this concentration seems to have continued to the year of 2022. To best support the energy transition, more funds need to flow to less mature technologies as well as to other sectors beyond electricity such as heating, cooling, and system integration.

Comparing renewables financing across countries and regions, the report shows that glaring disparities have increased significantly over the last six years. About 70% of the world's population, mostly residing in developing and emerging countries, received only 15% of global investments in 2020. Sub-Saharan Africa, for example, received less than 1.5% of the amount invested globally between 2000 and 2020. In 2021, investment per capita in Europe was 127 times that in sub-Saharan Africa, and 179 times more in North America.

The report emphasizes how lending to developing countries looking to deploy renewables must be reformed, and highlights the need for public financing to play a much stronger role, beyond mitigating investment risks. Recognizing the limited public funds available in the developing world, the report calls for stronger international collaboration, including a substantial increase in financial flows from the Global North to the Global South.

"For the energy transition to improve lives and livelihoods, governments and development partners need to ensure a more equitable flow of finance, by recognising the different contexts and needs," says IRENA Director-General, Francesco La Camera.

This joint report underscores the need to direct public funds to regions and countries with a lot of untapped renewables potential but find it difficult to attract investment. International cooperation must aim at directing these funds to enabling policy frameworks, the development of energy transition infrastructure, and to address persistent socio-economic gaps.

Achieving an energy transition in line with the 1.5°C Scenario also requires the redirection of US$0.7 trillion per year from fossil fuels to energy-transition-related technologies. But following a brief decline in 2020 due to COVID-19, fossil fuel investments are now on the rise. Some large multinational banks have even increased their investments in fossil fuels at an average of about US$0.75 trillion a year since the Paris Agreement.

In addition, the fossil fuel industry continues to benefit from subsidies, which doubled in 2021 across 51 countries. The phasing out of investments in fossil fuel assets should be coupled with the elimination of subsidies to level the playing field with renewables. However, the phaseout of subsidies needs to be accompanied by a proper safety net to ensure adequate standards of living for vulnerable populations. Barbara Buchner, CPI's Global Managing Director says,

> The path to net zero can only happen with a just and equitable energy transition. While our numbers show that there were record levels of investment for renewables last year, a greater scale-up is critically needed to avoid dangerous climate change, particularly in developing countries.

This is the third edition of the biannual joint report by IRENA and CPI. This report series analyses investment trends by technology, sector, region, source of finance, and financial instrument. It also analyses financing gaps, aiming to support informed policy making to deploy renewables at the scale needed to accelerate the energy transition. This third edition looks at the period of 2013–2020 and provides preliminary insights and figures for 2021 and 2022.

TUG OF WILL

While there are many avenues of coalition of the willing, there are also disconnects in many aspects of what needs to be done by whom, where, when, and how. Thus, there is often a "tug of will" among the participants, across the spectrum. This book aims to make a contribution by highlighting some of the essential goals, actions, and strategies. If we can turn a tug of will into a cohesive coalition of the willing, we will be further along the path of influencing our planet positively and mitigating the adverse impacts of climate change. The many paths of contributing to the climate change efforts include the following:

- Political
- Scientific
- Business
- Education
- Governmental

- Nongovernmental organizations (NGOs)
- Technological
- Engineering
- Mathematical modeling
- Administrative

According to IRENA (2023), global investments in energy transition technologies reached US$1.3 trillion in 2022, a record high. Yet the current pace of investment is not sufficient to put the world on track towards meeting climate or socio-economic development goals.

In 2022, global investments in energy transition technologies—renewable energy, energy efficiency, electrified transport and heat, energy storage, hydrogen and carbon capture and storage (CCS)—reached US$1.3 trillion despite the prevailing macro-economic, geopolitical and supply chain challenges. Global investments were up 19% from 2021 levels, and 50% from 2019, before the COVID-19 pandemic (Figure 2.1). This trend demonstrates a growing recognition of the climate crisis and energy security risks associated with over-reliance on fossil fuels.

Yet the current pace of investment is not sufficient; annual investments need to at least quadruple. Keeping the world on track to achieving the energy transition in line with the 1.5°C Scenario laid out in IRENA's *World Energy Transitions Outlook* 2022 will require annual investments of US$5.7 trillion on average between 2021 and 2030, and US$ 3.7 trillion between 2031 and 2050 (IRENA, 2022a).

Renewable energy investments for 2021 and 2022 represent preliminary estimates based on data from Bloomberg New Energy Finance (BNEF). As BNEF does not include large hydropower investments, these were estimated at US$7 billion per year, the annual average investment in 2019 and 2020. Energy efficiency data are from IRENA, 2023). These values are in constant 2019 dollars, while all other values are at current prices and exchange rates. Due to the lack of more granular data, the units could not be harmonized across the databases. For this reason, these numbers are presented together for indicative purposes only and should not be used to make comparisons between data sources. Data for other energy transition technologies come from BNEF (IRENA 2023).

Achieving an energy transition in line with the 1.5°C Scenario requires the redirection of US$0.7 trillion per year from fossil fuels to energy-transition-related technologies; but fossil fuel investments are still on the rise.

Fossil fuel investments had declined in 2020 (down 22% from the US$1 trillion invested in 2019) mainly due to the impacts of the COVID-19 pandemic on global energy markets (IEA, 2022c). Nevertheless, 2021 saw fossil fuel investments bounce back up 15% to US$897 billion (Figure 2.2), and preliminary data for 2022 suggest they might have almost returned to their pre-pandemic levels (+6%), reaching US$953 billion (IEA, 2022c).

Investment in energy is still going into funding new oil and gas fields instead of renewables and it is estimated that US$570 billion will be spent on new oil and gas development and exploration every year until 2030 (IISD, 2022).

Investors and banks have already committed to financing fossil fuel development over and above the limit needed to meet the 1.5°C target. Over the six years following

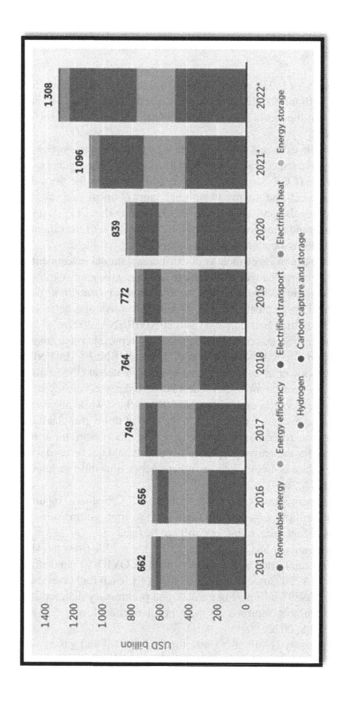

FIGURE 2.1 Annual global investment in renewable energy, energy efficiency and transition-related technologies, 2015–2022.

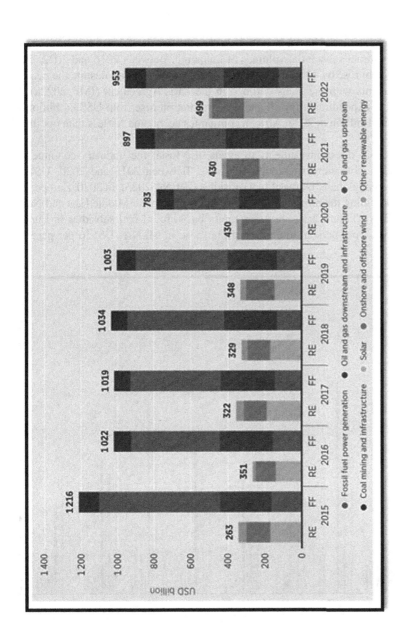

FIGURE 2.2 Annual investment in renewable energy vs. fossil fuels, 2015–2022. FF = fossil fuel; RE = renewable energy. Based on IRENA, 2023.

the Paris Climate Agreement, some large multinational banks maintained and even increased their investments in fossil fuels at an average of about US$750 billion dollars per year (IRENA, 2023). The world's 60 largest commercial banks invested around US$4.6 trillion in fossil fuels between 2015 and 2021, more than one-quarter of which came from US banks (IRENA 2023).

Fossil fuel companies based in emerging markets and developing economies have continued to attract substantial volumes of financing. Between 2016 and 2022, their outstanding debt rose by 400% for coal and 225% for oil and gas, despite the need to align investments with the goals outlined in the Paris Agreement (IMF, 2022a). In Africa, capital expenditures for oil and gas exploration rose from US$3.4 billion in 2020 to US$5.1 billion in 2022. African companies accounted for less than one-third of this sum.

In addition to direct investments in assets, the fossil fuel industry continues to receive considerable support through subsidies. Between 2013 and 2020, US$2.9 trillion was spent globally on fossil fuel subsidies (IRENA 2023). In 2020, Europe was the region providing the most subsidies, having overtaken the Middle East and North Africa (MENA) (Figure 2.3). On a per capita basis, fossil fuel subsidies in Europe totalled US$113 per person, more than triple those in MENA (US$36 per person).

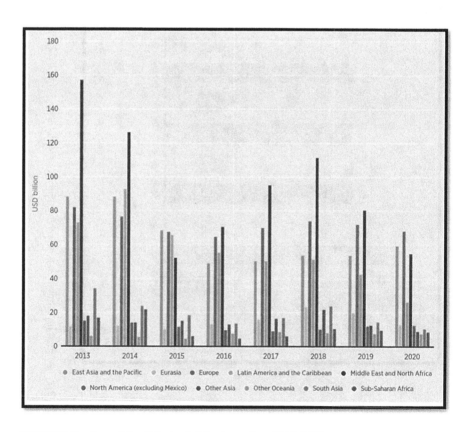

FIGURE 2.3   Annual fossil fuel subsidies by region, 2013–2020.

However, fossil fuel subsidies in MENA make up 1.56% of the gross domestic product (GDP) while in Europe, they constitute only 0.3% of GDP.

Subsidies doubled in 2021 across 51 countries, from US$362 billion in 2020 to US$697 billion, with consumption subsidies expected to have risen even further in 2022 due to contemporaneous price pressures (OECD and IEA, 2022). The phasing out of investments in fossil fuel assets should be coupled with the elimination of subsidies to ensure that the full costs of fossil fuels are reflected in their price and to level the playing field with renewables and other energy-transition-related technologies. However, the phaseout of subsidies needs to be accompanied by a proper safety net to ensure adequate standards of living for vulnerable populations (IRENA, 2022a).

Investments in renewable energy continue to grow, but not at the pace needed to achieve climate, energy access and energy security objectives along with other socioeconomic development goals by 2030.

Despite multiple economic, social and geopolitical challenges, annual investments in renewable energy continued a positive trend that began after 2018 (see Figure 2.4). Preliminary data suggest that in 2021, investments reached US$430 billion (24% up from 2020) and in 2022 they further increased by 16%, reaching almost US$0.5 trillion (ARENA, 2023). Yet, investment in 2022 was 40% of the average investment needed each year between 2021 and 2030 (about US$1.3 trillion in renewable power and the direct use of renewables) according to IRENA's 1.5°C Scenario.

Investments for 2021 and 2022 are preliminary estimates based on data from ARENA (2023). As BNEF data has limited coverage of large hydropower investments, these were assumed to be US$7 billion per year, equivalent to the annual average investment for the preceding two years. These figures represent "primary" financial transactions in both large- and small-scale projects that directly contribute to deployment of renewable energy, and therefore exclude secondary transactions, e.g., refinancing of existing debts or public trading in financial markets. Note that this is different from investments discussed in Chapter 3 of IRENA (2023) for the off-grid renewable energy sector which relates to corporate-level transactions (both primary and secondary) and is therefore different from investments discussed in Chapter 2 of IRENA (2023). For more details, please see the methodology document of IRENA (2023).

Investments are also not flowing at the pace or scale needed to achieve the improvements in livelihoods and welfare envisioned in the 2030 Agenda for Sustainable Development. Despite progress in energy access, approximately 733 million people had no access to electricity and nearly 2.4 billion people relied on traditional fuels and technologies for cooking at the end of 2020 (IEA, IRENA et al., 2022). Between 2010 and 2021, the off-grid renewables sector attracted more than US$3 billion (Wood Mackenzie, 2022a). Investments in off-grid solutions reached US$558 million in 2021, a 27% increase from 2020 (Figure 2.5). But this amount is far short of the US$2.3 billion needed annually in the sector between 2021 and 2030 to accelerate progress towards universal energy access (IRENA, 2023).

Although on the rise, off-grid investments are concentrated among seven large incumbent companies that have already reached scale and are looking to further solidify their market position through their ability to attract capital. The average transaction

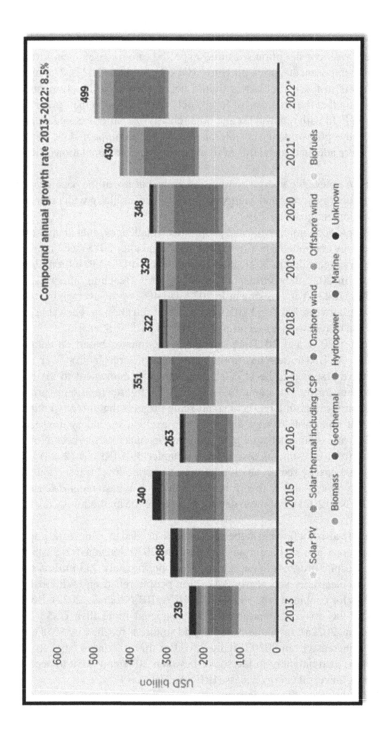

FIGURE 2.4 Annual financial commitments in renewable energy, by technology, 2013–2022. CAGR = compound annual growth rate; CSP = concentrated solar power; PV = photovoltaic.

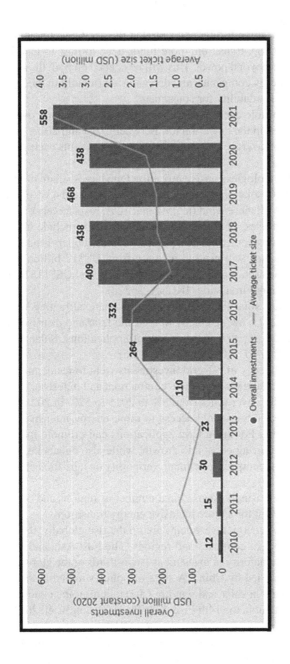

FIGURE 2.5   Annual investment in off-grid renewable energy and average transaction size, 2010–2021. Based on: Wood Mackenzie (2022a).

size climbed from US\$1.1 million in 2017 to US\$1.7 million by 2020, before more than doubling to US\$3.7 million in 2021 (Figure 2.5). While a trend of growing ticket size is a sign of sector growth and maturity, it may also indicate existing challenges for enterprises looking for smaller investments.

Investments have become further concentrated in specific technologies and uses. To best support the energy transition, more funds need to flow to less mature technologies and to sectors beyond power. This will be needed on both the supply side for off-grid renewable energy companies and demand side (mainly in the form of public funding) to enhance affordability for consumers.

While annual renewable energy investments have been growing over time, these have been concentrated in the power sector. Between 2013 and 2020, power generation assets attracted, on average, 90% of renewable investments each year, and up to 97% in 2021 and 2022.

Solar and wind technologies consistently attract the largest share of investment by a wide margin. In 2020, solar photovoltaic (PV) alone attracted 43% of the total, followed by onshore and offshore wind (at 35% and 12%, respectively).

Investments in end uses, i.e. direct applications, which include heat generation (e.g. solar water heaters, geothermal heat pumps, biomass boilers) and transport (e.g. biofuels) are lagging; they will need to increase from US\$17 billion in 2020 to an average US\$284 billion each year between now and 2030 and US\$115 billion through 2050 to achieve the energy transition (IRENA, 2022a).

In the off-grid space, solar PV products also dominate, attracting 92% of overall investments in 2010–2021, owing chiefly to their modular and distributed characteristics, and their adaptability to a wide variety of applications. Solar home systems (SHSs) are the most funded technology (Figure 2.6).

Even though the majority of off-grid investments went to residential applications between 2010 and 2021, the share going to commercial and industrial (C&I) applications has been expanding over time (from 8% in 2015 to 32% in 2021) as consumer needs grow beyond basic household access to more energy-intensive uses in local industry and agriculture. Powering C&I applications can promote local economies by creating jobs and spurring economic growth, while also enhancing food security and resilience against the impacts of climate variability on agri-food chains (IRENA, 2016b).

Investments are increasingly focused in a number of regions and countries. They need to be more universal for a more inclusive energy transition.

Although renewable energy investments are on the rise globally, they continue to be focused in a number of countries and regions. The East Asia and Pacific region continues to attract the majority of investment—two-thirds of the global total in 2022 (Figure 2.7)—primarily led by China. A suite of policies including tax exemptions have driven investments in solar and wind in China, putting the country on track to meeting the targets set out in the 14th Five-Year Plan (Carbon Brief, 2021). Viet Nam saw investment in solar PV grow by an average of 219% per year between 2013 and 2020, driven mainly by feed-in tariffs (Lorimer, 2021). North America excluding Mexico attracted the second-largest share of investment in 2022, mainly driven by the production tax credit in the United States, followed by Europe, where net-zero

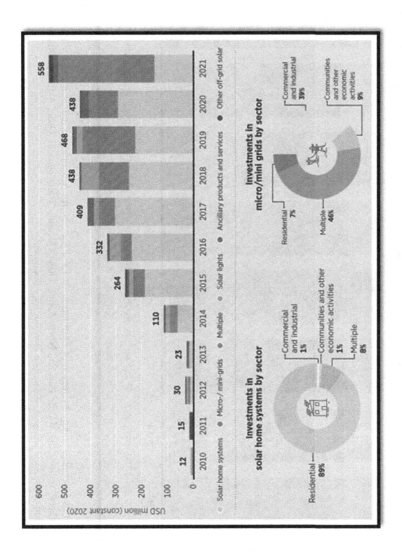

FIGURE 2.6 Annual investment in off-grid renewable energy, by off-grid product, and energy use, 2010–2021. Based on: Wood Mackenzie (2022a).

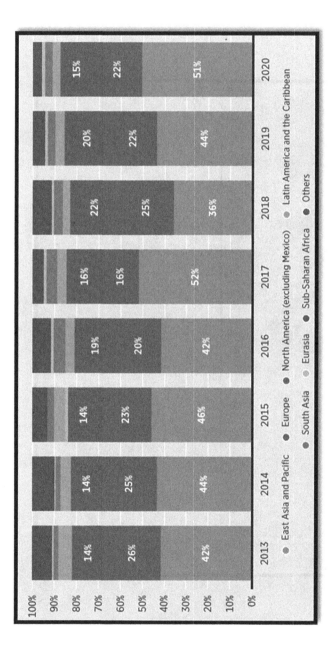

FIGURE 2.7 Investment in renewable energy by region of destination, 2013–2022. "North America (excluding Mexico)" includes Bermuda, Canada and the United States. "Others" include the Middle East and North Africa, Other Oceania, Transregional, Other Asia and Unknown. For more details on the geographic classification used in the analysis, please see methodology document.

commitments and extensive policies to phase out fossil fuels are driving growth in renewables.

In the United States, the 2022 Inflation Reduction Act—encompassing new tax credits, US$30 billion in grants and loans for clean energy generation and storage, and US$60 billion in support of manufacturing of low-carbon components—is expected to attract US$114 billion investment by 2031. In Europe, the European Commission presented a *Green Deal Industrial Plan for the Net-Zero Age*, which would provide investment aid and tax breaks towards technological development, manufacturing, production and installation of net-zero products in green sectors including renewables and hydrogen (Bloomberg, 2023; IRENA, 2023). The plan looks to mobilize EUR 225 billion in loans from its existing Recovery and Resilience Facility, and an additional EUR 20 billion in grants (European Commission, 2023).

Regions home to about 120 developing and emerging markets continue to receive comparatively low investment. Across these regions, the bulk of renewable energy investments is captured by a handful of countries: Brazil, Chile, and India. In other words, more than 70% of the world's population, mostly residing in developing and emerging countries, received only 15% of global investments in renewables in 2022. Further, the share of renewable energy investments going to these regions has been progressively declining year on year (e.g. from 27% in 2017 to 15% in 2020). In absolute terms, annual investments have been declining precipitously since 2018 at an average rate of 36%. Countries defined as "least developed" by the Intergovernmental Panel on Climate Change attracted only 0.84% of renewable energy investments on average between 2013 and 2020.

Looking at investments on a per capita basis further reveals the disparity in investments. In East Asia and Pacific, investment per capita increased by 19% between 2015 and 2021 from US$88/person in 2015 to US$105/person in 2021. The bulk of the increase took place in China, and, in fact, the region excluding China experienced a decrease of 20%. In South Asia, investments per capita declined by 26% between 2015 and 2021; however, the true extent of the decline is masked by India which saw investment per capita grow by 34% in the same period. Excluding India, investment per capita declined by 76% from US$20/person in 2015 to US$5/person in 2021. The most striking—and rapidly growing—disparity is between sub-Saharan Africa and both North America (excluding Mexico) and Europe. In 2015, renewable energy investment per capita in North America (excluding Mexico) or Europe was just about 22 times higher than that of sub-Saharan Africa. In 2021, investment per capita in Europe was 127 times that in sub-Saharan Africa (which in 2021 fell to just US$1/person from US$7/person in 2015), and North America was 179 times more.

Sub-Saharan Africa remains the primary destination for investment in off-grid renewables. The region attracted US$2.2 billion in 2010–2021—more than 70% of global off-grid investments. Electrification rates are among the lowest in the world, with 568 million people lacking access to electricity in 2020 (IRENA, 2023). Within sub-Saharan Africa, East Africa—home to three of the top five recipient countries of off-grid investment (Kenya, the United Republic of Tanzania and Rwanda)—attracted 43% of the total. Investment in these destinations benefited from the existing mobile money ecosystem, which was leveraged by the pay-as-you-go (PAYG) business

48 Systems Engineering

model. Approximately 78% of the total commitments in off-grid renewables in 2010–2021 (or US$2.4 billion) involved the funding of companies or projects using PAYG, with East Africa accounting for US$917 million.

During the COVID-19 pandemic, off-grid renewable energy investments in Southeast Asia declined by 98%, leaving key off-grid markets even more vulnerable. Although the majority of countries in the region have achieved high or near-universal rates of electricity access, parts of the populations in countries such as Myanmar and Cambodia (26% and 15%, respectively in 2020) still lack access to electricity (World Bank, 2022). Whereas the region attracted US$137 million in off-grid renewable energy investments over 2018–2019 (led primarily by Myanmar), during 2020–2021, investments plummeted to US$3 million, likely due to the impacts of the COVID-19 pandemic and political developments (IRENA, 2023).

Investments have been primarily made by private actors. Private capital flows to the technologies and countries with the least risks—real or perceived. The private sector provides the lion's share of global investments in renewable energy, committing around 75% of the total in the period 2013–2020 (Figure 2.8). The share of public versus private investments varies by context and technology. Typically, lower shares of public finance are devoted to renewable energy technologies that are commercially viable and highly competitive, which makes them attractive for private investors. For example, in 2020, 83% of commitments in solar PV came from private finance. Meanwhile, geothermal and hydropower rely mostly on public finance; only 32% and 3% of investments in these technologies, respectively, came from private investors in 2020.

Globally, commercial financial institutions and corporations are the main private finance providers, accounting together for almost 85% of private finance for renewables in 2020 (Figure 2.9). Up until 2018, private investments came predominantly from corporations (on average, 65% during 2013–2018), but in 2019 and 2020 the share of corporations went down to 41% per year, and a larger share of investments was filled by commercial financial institutions (43%).

This aligns with the falling share of equity financing globally, from 77% in 2013 to 43% by 2020 (Figure 2.10) as corporations together with households/individuals provided 83% of equity financing during 2013–2020 (Figure 2.8). During this time, the share of debt financing increased from 23% in 2013 to 56% in 2020 (Figure 2.10). This is likely linked to the maturation and consolidation of major renewable technologies such as solar PV and onshore wind that are able to attract high levels of debt, as lenders are able to envision regular and predictable cash flows over the long term, facilitated by power purchase agreements (PPAs) in many countries.

In the off-grid space, debt and equity investments contributed about 47% and 48% of the overall financing, respectively between 2010 and 2021, with an additional 5% contributed by grants. By technology, debt financing constituted the majority of the investments in solar home systems and solar lights (54% of the total and rising over time) while equity financing dominated the micro-/mini-grid space. Prior to the COVID-19 pandemic, the majority of off-grid financing came from equity investments owing to the domination by private equity, venture capital and infrastructure funds and the lack of debt access for the sector. Ever since, the share of private equity has seen a relative decline (Figure 2.11), in part due to the uncertainties posed by

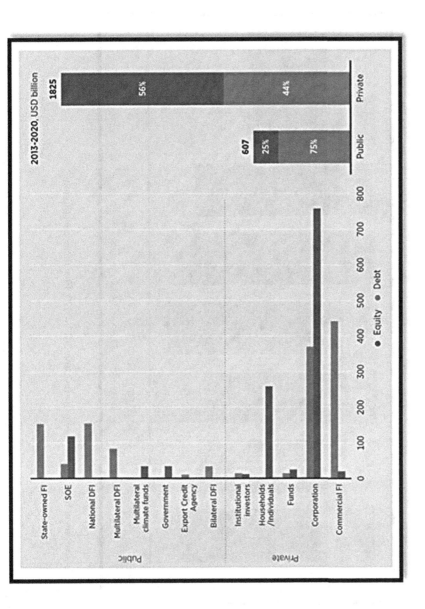

FIGURE 2.8 Debt and equity investment by type of investor, 2013–2020. DFI = development finance institution; FI = finance institution; SOE = state-owned enterprise.

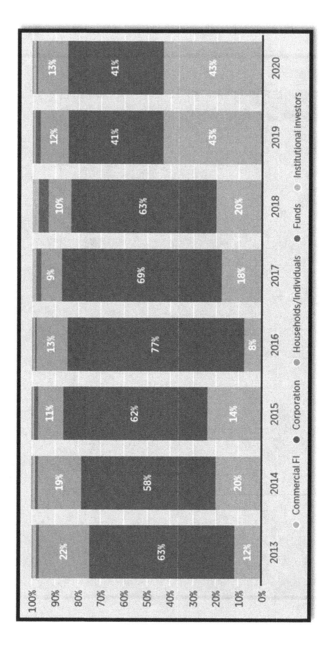

FIGURE 2.9 Private investment in renewable energy by investor, 2013–2020. FI = finance institution.

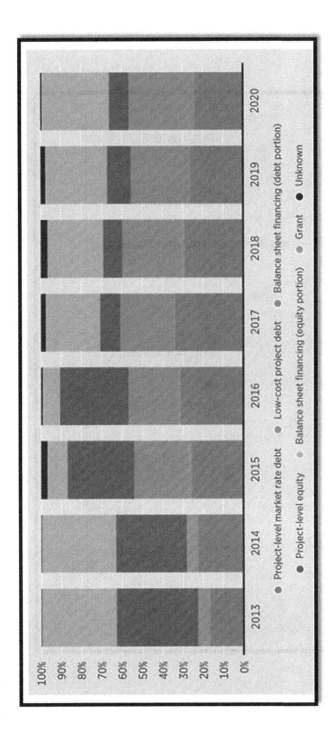

FIGURE 2.10 Investment in renewable energy, by financial instrument, 2013–2020.

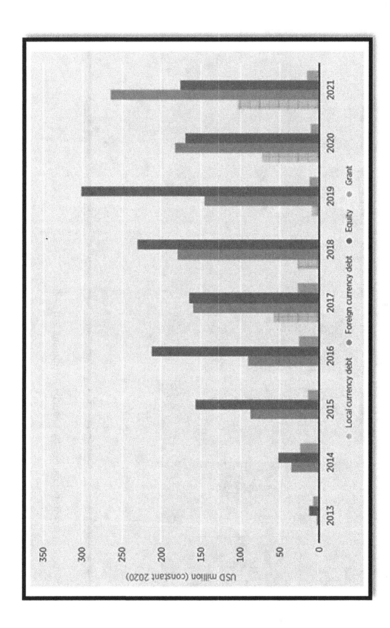

FIGURE 2.11 Annual investment in off-grid renewable energy, by financing instrument and local versus foreign currency debt, 2013–2021. Based on: Wood Mackenzie (2022a).

the pandemic, and the limited track record of exits and capital recycling in the sector. The contribution of debt has increased sharply over the past two years, particularly as debt-preferring DFIs bolstered their support during the pandemic (Figure 2.15) and major off-grid companies were able to capitalize on their strong market position to secure (large-size) predominantly debt-based deals from both public and private investors (IRENA, 2023). Another remarkable trend is the increase in local currency debt, driven mainly by markets in Kenya and Nigeria.

Going forward, widespread mobilization of low-cost debt will be critical for the deployment of capital-intensive renewable energy projects, while equity financing will also remain key, particularly to kick-start relatively less mature technologies, and finance projects in relatively high-risk or credit-constrained contexts.

The majority of public investments are made domestically with relatively little international collaboration. The international flow of public money to renewable energy has been in decline since 2018.

Public funds are limited, so governments have been focusing what is available on de-risking projects and improving their risk–return profiles to attract private capital.

Globally, the public sector provided less than one-third of renewable energy investment in 2020. State-owned financial institutions, national DFIs, and state-owned enterprises were the main sources that year, providing more than 80% of public finance (Figure 2.12). Multilateral DFIs provided 9% of public finance—in line with their past annual commitments—and accounted for about half of international flows coming from the public sector. Commitments from bilateral DFIs in 2020 fell 70% compared to 2019, largely due to a 96% decline in international commitments by the German Development Bank (KfW). This means that multilateral and bilateral DFIs provided less than 3% of total renewable energy investments in 2020.

In addition, financing from DFIs was provided mainly in the form of debt financing at market rates (requiring repayment with interest rates charged at market value). Grants and concessional loans amounted to just 1% of total renewable energy finance, equivalent to USD 5 billion. Since the interest rates are the same, the only difference that DFI financing provides is to making finance available, albeit at the same high costs for users. Figure 2.13 illustrates the portion of DFI funding provided in the form of grants and low-cost debt.

In the off-grid space, the role of the public sector, in particular DFIs, is much more important. DFIs were the largest public capital providers (accounting for 79% of the public investments in off-grid solutions and 27% of the total investments in off-grid solutions in 2010–2021). Notably, DFIs' contributions after the pandemic constitute half of their overall contributions since 2010 (Figure 2.14).

Public finance flows to the Global South are essential to achieving the 1.5°C Scenario and its socio-economic benefits (together with progressive fiscal measures and other government programmes such as distributional policy, as outlined in IRENA [2022a]). In fact, almost 80% of the off-grid investments between 2010 and 2021 involved North–South flows. However, the international flow of public finance going to renewable energy in the broader context has been in decline since 2018 (IRENA, 20232). Preliminary data show that the downtrend continued through 2021.

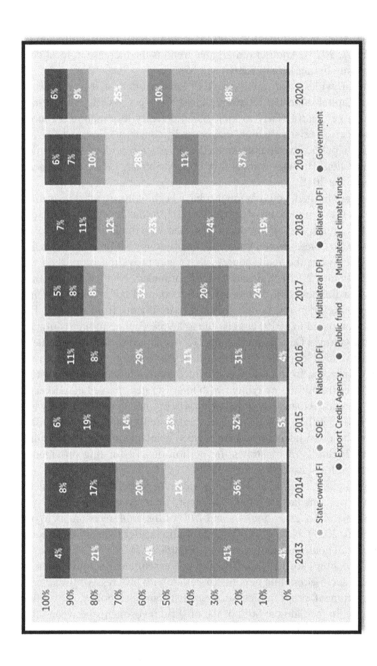

FIGURE 2.12 Public investment in renewable energy by investor type, 2013–2020. DFI = development finance institution; FI = finance institution; SOE = state-owned enterprise.

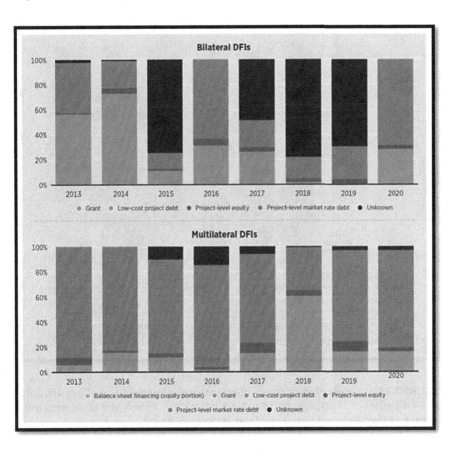

FIGURE 2.13   Portion of DFI funding in the form of grants and low-cost debt. DFI = development finance institution.

To achieve a just and inclusive energy transition, public financing—including through international collaboration—has a critical role to play across a broad spectrum of policies.

Among risk mitigation instruments, sovereign guarantees have been preferred for lenders looking to obtain a "one-size-fits-all" solution for credit risks. But such guarantees are treated as contingent liabilities and may hamper a country's ability to take on additional debt for critical infrastructure development and other investments (IRENA, 2020a). Moreover, sovereign debts are already stressed to their breaking point in many emerging economies grappling with high inflation and currency fluctuations or devaluations in the wake of the COVID-19 pandemic. In this macroeconomic environment, many countries cannot access affordable capital in international financial markets or provide sovereign guarantees to mitigate risk.

Given the urgent need to step up the pace and geographic spread of the energy transition, and to capture its full potential in achieving socio-economic development goals, more innovative instruments are needed that help under-invested countries

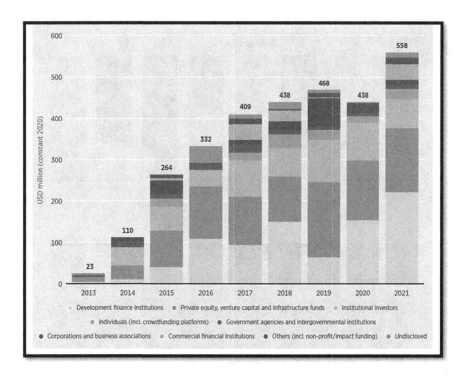

FIGURE 2.14   Annual commitments to off-grid renewable energy by type of investor, 2015–2021. Definitions of all investor type included in this analysis are provided in the accompanying methodology document. Based on: Wood Mackenzie (2022a).

reap the long-term benefits of the energy transition without putting their fiscally constrained economies at a further disadvantage.

Public funding must flow into the renewable energy sector (covering all segments of the value chain), the wider energy sector and the economy as a whole, for a just and equitable energy transition. Public funds can be mobilized and provided using a variety of instruments. Figure 2.15 shows the types of instruments that can be used to channel public finance, the sources of public funds (domestic or international through collaboration) and the intermediaries that can help channel them (e.g. governments, national DFIs, local banks, multilateral and bilateral DFIs, export credit agencies, global funds including the Just Energy Transition Partnership [JETP] and UN-linked funds such as the Green Climate Fund).

These instruments can be existing or newly designed and may include: (1) government spending such as grants, rebates, and subsidies; (2) debt including existing and new issuances, credit instruments, concessional financing and guarantees; (3) equity and direct ownership of assets (such as transmission lines or land to build projects); and (4) fiscal policy and regulations including taxes and levies, exemptions, accelerated depreciation, deferrals and regulations such as PPAs (especially when the tariffs paid to producers—in addition to the cost of running the system—are lower than what is collected by consumers and the difference is paid through a government subsidy).

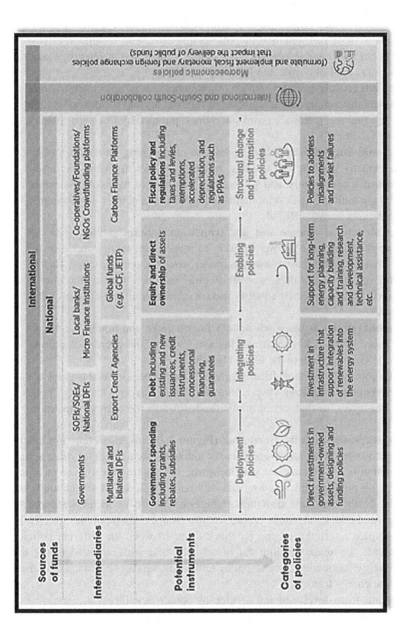

FIGURE 2.15   The flow of public finance for a just and inclusive energy transition. DFI = development finance institution; GCF = Green Climate Fund; JETP = Just Energy Transition Partnership; NGO = non-governmental organization; PPA = power purchase agreement; SOFI = state-owned financial institution; SOE = state-owned enterprise.

As shown in Figure 2.15, public finance flows via instruments in various policy categories of IRENA's broad policy framework. Examples include the following:

1. Under **deployment policies**, public funds can flow as direct investments in government-owned energy-transition-related assets, public–private partnerships, or in designing and funding policies that can attract or support private investment (e.g. capital subsidies, grants and tariff-based mechanisms such as auctions, feed-in tariffs, and feed-in premiums).

2. Under **integrating policies**, public investments can go into infrastructure and assets that support the integration of renewables into the energy system (e.g. regional and national transmission lines, pumped hydroelectric energy storage facilities).

3. Under **enabling policies**, public money can support long-term energy planning, capacity building and training, research and development, the development of local industry and value chains, as well as technical assistance offered via multilateral development banks (MDBs) and inter-governmental organizations such as IRENA.

4. Under **structural change and just transition policies**, public funds can go into the redesign of power markets to make them more conducive for large shares of variable renewable energy, towards compensation for the phasing-out of fossil fuels, as well as policies to ensure that the energy transition promotes gender equality and social inclusion, among many other priorities.

5. The **global policy framework** defines international and South–South collaboration, which is key to structuring and ensuring the international flows from the Global North to the Global South.

6. In addition, although not directly related to any specific sector, there are **macroeconomic policies** (fiscal, monetary, and currency exchange policies) that affect the delivery of public funds towards the energy transition.

Some elements presented in the framework (Figure 2.15) might overlap. For example, tax incentives are at the same time fiscal or macroeconomic policies while acting as deployment policies, and funding grid infrastructure can be viewed as an enabling or an integrating policy. While funding capacity building is part of an enabling policy, these funds also facilitate structural change, being part of social development programmes, and education, social protection and compensation policies, etc. Thus, there are complex interlinkages and feedback loops between the different policies and instruments. By understanding the broad structural workings underlying the renewable energy "economy", public policy and financing can be strategically used to advance the energy transition.

Governments from developed and developing countries will play a central role in providing an enabling environment for both public and private investments.

A more comprehensive way of defining risk (including risk sharing) is needed. A narrow investor-centric focus on the risk of investment in energy assets not paying off needs to be broadened to include environmental, planetary, and social risks. These include the risk of leaving a large part of the population out of the energy transition and locked in underdevelopment, and the risk of the Sustainable Development Goals remaining far from being met. This is how investment risks must be viewed from the perspective of governments and the international community. And with the very limited public funds available in the developing world, the international community must step up.

The availability of capital for public investments in renewable energy will need to be increased, and lending to developing nations transformed. Today's environment calls for a fundamental shift in how lending is made to developing nations, especially those affected by economic and climate crises, and particularly how countries in the Global North support countries in the Global South to cope with and adapt to crises related to climate change, the cost of living and debt. The situation in developing countries is being made more difficult amid tightening monetary policies and a strengthening US dollar. One in five countries is experiencing fiscal and financial stress, which left unaddressed would deepen hardship, increase debt defaults, widen inequality and delay the energy transition.

At the 27th United Nations Climate Change Conference (COP27) a decision was reached to establish a loss and damage fund, particularly for those nations most vulnerable to climate events. Details regarding the amounts involved, and how the facility will be set up and operationalized are yet to be negotiated. The fund is expected to address adverse effects of climate impacts such as droughts, floods, rising seas and other disasters that impair the deployment of renewable energy.

Tapping pools of public funds for both developed and developing countries without burdening the fiscal space remains a key priority. Governments should adopt a "doing more with what is available" approach through enhanced collaboration among DFIs and MDBs, and by exploring the following mechanisms:

Capital release from balance sheets of DFIs. Balance sheets of investors and financial institutions disclose rights and obligations connected to the owning and lending of assets. It is possible for DFIs to use those elements to raise additional funds through posting existing assets as collateral (provided their value is free and clear of any encumbrances), and partially repackaging receivables from guaranteed loan repayments (e.g. loans that are guaranteed by insurers) into new financial structured products in the market. The DFIs could offer a (high-rated) new debt product and traded on international exchanges. However, such a product should be used with rigorous due diligence. Collateralized debt obligations are asset-backed securities that bundle together a diversified portfolio of instruments (e.g. loans, bonds). Cash flows from underlying assets are used to repay investors.

Product Innovation among MDBs

Multilaterals benefit from the convening power granted by shareholders in both developed and developing countries, to craft, implement and operate innovative frameworks to mobilize capital and mitigate risks. In particular, liquidity facilities can be scaled up to assist renewable energy investors in fulfilling their business obligations by ensuring an uninterrupted flow of payments from off-takers—without posing a burden on the fiscal space of developing countries (local-currency-denominated PPAs can also benefit from this facility). These liquidity facilities can evolve to incorporate the role of guarantor supported by MDBs and DFIs in compliance with guidelines issued by multilaterals and agreed by shareholders. The highly capitalized guarantor becomes a supranational facility to mitigate credit and foreign exchange risks for renewable energy investors and lenders. MDBs, under the approval of host governments, can allocate funds and credit lines to the facility up to prudent limits determined by ministries of finance and central banks.

BROADENING CAPITALIZATION ROUTES FOR MDBS

Capital calling from shareholders has been the common approach adopted by multilaterals to expand technical assistance and lending programmes. The new capital increases MDBs' fund availability and enables them to place bonds in the global capital market, thereby raising additional capital. Bonds are placed as AAA-rated obligations guaranteed by MDBs—de facto, such institutions have an enviable track record recognised by countries and market participants in managing risks—that can be placed in the market, if appropriate financing vehicles are used and target markets are identified. MDBs should now consider risk-tiered debt obligation placements with a different investment grade (BBB+ and above, e.g. multi-rated green bonds), implying different level of returns to bondholders. The initiative broadens access to the investor base—from institutional investors and sovereign wealth funds to corporate/qualified investors—increasing the amount of capital that could become available and deployed in renewable energy investments.

Meanwhile, public finance and policy should continue to be used to crowd in private capital. Policies and instruments beyond those used to mitigate risks are needed.

Public finance should continue to be used strategically to crowd in additional private capital. Risk mitigation instruments (e.g. guarantees, currency hedging instruments, and liquidity reserve facilities) will still play a major role, but public finance and policy must go beyond risk mitigation. Examples include funding capacity building, support for pilot projects and innovative financing instruments such as blended finance initiatives, etc. In addition, policy-makers may consider the following:

> Incentivize an investment swap from fossil fuels to renewable energy by banks and national oil companies. Incentivising investors to divert funds towards the energy transition can be done through measures such as phasing out of fossil fuel subsidies and adapting fiscal systems to account for the environmental, social and health impacts of a fossil-fuel-based energy system. However, the phaseout of subsidies should be accompanied by a proper safety net to ensure adequate standards of living for vulnerable populations.
>
> (**IRENA**, 2022a)

A supplemental way of incentivizing this shift is through highlighting and recognising the leadership role of those institutions that are paving the road through early investments in the energy transition. More than 30 significant financial institutions, including banks, insurers, asset owners, and asset managers, have committed to stop financing fossil fuels. Governments and civil societies can take action to reward their leadership and encourage other institutions to take similar steps. After that, public pressure, along with policy and regulation, can further influence financial decision-making in favour of renewable energy and other energy transition technologies (IRENA 2023).

MOBILIZE INSTITUTIONAL INVESTMENT AND PROMOTE GREATER USE OF GREEN BONDS FOR RENEWABLES

With about US$87 trillion of assets under management, institutional investors have a key role to play in reaching the investment levels required for the ongoing global energy transition. Greater participation of institutional capital will require a combination of effective policies and regulations, capital market solutions that address the needs of this investor class (e.g. green bonds), as well as a variety of internal changes and capacity building on the part of institutional investors (IRENA, 2020d).

Green bonds can help attract institutional investors and channel considerable additional private capital in the renewable energy sector, helping to fill the significant outstanding investment gap. Green bonds have experienced significant growth over the past decade (about 103% a year in 2011–2021), increasing from about US$800 million of issuances in 2007 to about US$545 billion of issuances in 2021—an all-time annual high despite pandemic-induced economic challenges. The cumulative value of green bond issuances broke the US$1 trillion threshold at the end of 2020 and stood at about US$1.64 trillion as of the end of 2021 (IRENA, 2023). Some recommended actions for policy makers and public finance providers to further increase green bond issuances include the adoption of green bond standards in line with international climate objectives, the provision of technical assistance and economic incentives for green bond market development and the creation of bankable project pipelines (IRENA, 2023).

IMPLEMENT REGULATORY SANDBOXES FOR BROADENING ACCESS TO CAPITAL AND CREDIT INSTRUMENTS

Regulatory sandboxes designed to serve broader social and environmental goals can help unlock more investments. By enacting regulatory sandboxes for start-ups and investors for both grid and off-grid initiatives, new solutions may emerge towards enabling access to pools of capital/credit instruments. Such initiatives can benefit from MDBs' support (IRENA, 2023) in connection with other available funding agencies at local, regional and global levels. Furthermore, companies can be invited to participate in the sandbox with a view to pilot innovative concepts that facilitate risk mitigation, including foreign exchange risks in electricity exchanges.

FACILITATE LOCAL CURRENCY LENDING AND DENOMINATE PPAS (AT LEAST PARTIALLY) IN LOCAL CURRENCIES

Local currency PPAs are helpful to address the risks of currency devaluations which may otherwise cripple power off-takers' ability to make payments to power producers in hard currency (such as the USD) at times when the domestic currency plummets. Relatively established markets in the off-grid space, for instance, such as Kenya and Nigeria are seeing more local currency debt financing.

During 2020–2021, about 28% of debt in the two countries was denominated in local currencies (primarily the Kenyan shilling, followed by the Nigerian naira), compared with just 11% during the pre-pandemic years. Going forward, low-cost local currency financing will be preferred for the next phase of the off-grid renewable energy sector's development. A complementary mechanism to address foreign currency risks is to facilitate local currency lending for projects with development capital channeled through intermediaries, including national banks or non-banking financial institutions. Several countries, including Bangladesh, Brazil and Jordan, have piloted such approaches to catalyse investment into the renewable energy sector.

ENHANCE THE PARTICIPATION OF CORPORATE ACTORS

Although companies that produce renewable energy are already providing substantial investment in the sector, non-energy-producing corporations have a preeminent role to play in the energy transition by driving demand for renewable energy. By setting up the right enabling framework, policy makers can encourage active corporate sourcing and unlock additional capital in the sector. Recommended actions include, for example, establishing a transparent system for the certification and tracking of renewable energy attribute certificates, enabling third-party sales between companies and independent power producers, and creating incentives for utilities to provide green procurement options for companies (IRENA, 2018b).

INCENTIVIZE THE PARTICIPATION OF PHILANTHROPIES

According to Oxfam's report titled *Survival of the Richest: How We Must Tax the Super-Rich Now to Fight Inequality*, the richest 1% own almost half of the world's wealth while the poorest half of the world own just 0.75% (Oxfam, 2023). To tap into the existing wealth, governments should look at incentivizing philanthropies to mobilize additional funds into support for renewable energy that can help fight poverty, inequality, climate change and humanitarian crises. Philanthropies are playing an increasingly important role in bridging funding gaps, especially in the energy access context, where funds have gone into market development (e.g. technology innovation funds) and delivering financing for end users and enterprises through various instruments, such as results-based grants and equity. Individuals (high-net-worth individuals, families or households) invested an average of US$20 million per year in off-grid renewables during 2015–2021, primarily through dedicated crowdfunding platforms (IRENA, 2022f). In 2021, individuals, bequests, foundations and corporations gave an estimated US$485 billion to charities in the United States alone. These were distributed towards education, human services, foundations, public-society benefit organizations, health, international affairs, and environmental and other social services (Giving USA 2022). The energy transition being tied to all these objectives, tapping into these funds can help fill gaps left by governments, and support the livelihoods and well-being of relatively poor populations without relying on fossil fuels (IRENA, 2023).

CONCLUSIONS

The current crises present both a challenge and an opportunity for accelerated deployment of renewable energy. On the one hand, they provide the political momentum for hurrying the deployment of energy transition technologies. Indeed, the crisis in Ukraine has forced a global reckoning with the fact that 80% of the world's population live in countries that are net energy importers, a situation that carries profound implications for energy security and affordability.

On the other hand, tighter fiscal circumstances and higher costs of capital are dimming the prospects for renewable energy, which is particularly capital-intensive. The current uncertain macroeconomic outlook, with inflation at levels not seen in many countries for over 40 years and the possibility of sovereign defaults, could threaten renewable energy development, especially in low-income countries.

REFERENCES

Gunfaus, M. T. et al. (2022), "Just Energy Transition Partnerships: Can They Really Make a Difference, and How?", IDDRI, www.iddri.org/en/publications-and-events/blog-post/just-energy-transition-partnerships-can-they-really-make (accessed 13 February 2023).

GWEC (2022), *Global Wind Report 2022*, Global Wind Energy Council, Brussels, https://gwec.net/global-wind-report-2022/.

Hadley, S. (2022), "What's the State of Play on Just Energy Transition Partnerships?", ODI: Think Change, www.odi.org/en/insights/whats-the-state-of-play-on-just-energy-transition-partnerships/ (accessed 13 February 2023).

IEA (2020), "World Energy Investment 2020", International Energy Agency, Paris, https://iea.blob.core.windows.net/assets/ef8ffa01-9958-49f5-9b3b-7842e30f6177/WEI2020.pdf.

IEA (2021a), "Global Hydrogen Review 2021", International Energy Agency, Paris, https://iea.blob.core.windows.net/assets/e57fd1ee-aac7-494d-a351-f2a4024909b4/GlobalHydrogenReview2021.pdf.

IEA (2021b), "Renewables 2021: Analysis and Forecast to 2026", International Energy Agency, Paris, https://iea.blob.core.windows.net/assets/5ae32253-7409-4f9a-a91d-1493ffb9777a/Renewables2021-Analysisandforecastto2026.pdf.

IEA (2021c), "The Cost of Capital in Clean Energy Transitions – Analysis", International Energy Agency, www.iea.org/articles/the-cost-of-capital-in-clean-energy-transitions (accessed 31 January 2023).

IEA (2021d), "Hydropower Special Market Report: Analysis and Forecast to 2026", International Energy Agency, Paris, https://iea.blob.core.windows.net/assets/4d2d4365-08c6-4171-9ea28549fabd1c8d/HydropowerSpecialMarketReport_corr.pdf.

IEA (2022a), "Energy Efficiency Investment, 2015–2021", International Energy Agency, www.iea.org/data-and-statistics/charts/energy-efficiency-investment-2015-2021.

IEA (2022b), "World Energy Investment 2022 Datafile—Data Product", International Energy Agency, www.iea.org/data-and-statistics/data-product/world-energy-investment-2022-datafile (accessed 3 January 2023).

IEA (2022c), "World Energy Investment 2022", International Energy Agency, Paris, www.iea.org/reports/world-energy-investment-2022 (accessed 3 January 2023).

IEA, International Renewable Energy Agency, United Nations Statistics Division, World Health Organization and World Bank (2022), "Tracking SDG7: The Energy Progress Report 2022", World Bank, Washington, DC, www.irena.org/-/media/Files/IRENA/Agency/Publication/2022/Jun/SDG7_Tracking_Progress_2022.pdf.

IISD (2022), "Navigating Energy Transitions: Mapping the Road to 1.5°C, International Institute for Sustainable Development", Winnipeg, Manitoba, Canada, www.iisd.org/publications/report/navigating-energy-transitions (accessed 4 January 2023).

IMF (2022a), "Global Financial Stability Report", International Monetary Fund, Washington, DC, www.imf.org/en/Publications/GFSR/Issues/2022/10/11/global-financial-stability-reportoctober-2022 (accessed 30 January 2023).

IMF (2022b), "IMF Annual Report of the Executive Board", International Monetary Fund, Washington, DC, www.imf.org/external/pubs/ft/ar/2022/downloads/imf-annual-report-2022-english.pdf.

ING (2022), "How the Ukraine War Has Affected Asia's Race to Net Zero", *Internationale Nederlanden Groep*, https://think.ing.com/bundles/how-the-ukraine-war-has-affected-asias-race-to-net-zero.

IRENA (2016a), *Unlocking Renewable Energy Investment: The Role of Risk Mitigation and Structured Finance*, International Renewable Energy Agency, Abu Dhabi, www.irena.org/publications/2016/Jun/Unlocking-Renewable-Energy-Investment-The-role-of-risk-mitigation-and-structured-finance.

IRENA (2016b), *Renewable Energy Benefits: Decentralised Solutions in the Agri-Food Chain*, International Renewable Energy Agency, Abu Dhabi, www.irena.org/publications/2016/Sep/Renewable-Energy-Benefits-Decentralised-solutions-in-agri-food-chain.

IRENA (2017), *Geothermal Power: Technology Brief*, International Renewable Energy Agency, Abu Dhabi, www.irena.org/publications/2017/Aug/Geothermal-power-Technology-brief.

IRENA (2018a), *Off-grid Renewable Energy Solutions: Global and Regional Status and Trends*, International Renewable Energy Agency, Abu Dhabi, https://irena.org/publications/2018/Jul/Off-grid-Renewable-Energy-Solutions.

IRENA (2018b), *Corporate Sourcing of Renewable Energy: Market and Industry Trends*, International Renewable Energy Agency, Abu Dhabi, www.irena.org/Publications/2018/May/Corporate-Sourcing-of-Renewable-Energy.

IRENA (2019), *Hydrogen: A Renewable Energy Perspective*, International Renewable Energy Agency, Abu Dhabi., www.irena.org/publications/2019/Sep/Hydrogen-A-renewable-energy-perspective.

IRENA (2020a), *Renewable Energy Finance: Sovereign Guarantees*, International Renewable Energy Agency, Abu Dhabi, https://irena.org/-/media/Files/IRENA/Agency/Publication/2020/Jan/IRENA_RE_Sovereign_guarantees_2020.pdf.

IRENA (2020b), *Fostering a Blue Economy: Offshore Renewable Energy*, International Renewable Energy Agency, Abu Dhabi, www.irena.org/-/media/Files/IRENA/Agency/Publication/2020/Dec/IRENA_Fostering_Blue_Economy_2020.pdf

IRENA (2020c), *Mobilising Institutional Capital for Renewable Energy*, International Renewable Energy Agency, Abu Dhabi, www.irena.org/publications/2020/Nov/Mobilising-institutional-capital-for-renewable-energy.

IRENA (2020d), *The Post-COVID Recovery: An Agenda for Resilience, Development and Equality*, International Renewable Energy Agency, Abu Dhabi, www.irena.org/-/media/Files/IRENA/Agency/Publication/2020/Jun/IRENA_Post-COVID_Recovery_2020.pdf.

IRENA (2022a), *World Energy Transitions Outlook: 1.5°C Pathway*, International Renewable Energy Agency, Abu Dhabi, www.irena.org/publications/2022/Mar/World-Energy-Transitions-Outlook-2022 (accessed 10 August 2022).

IRENA (2022b), *Renewable Energy Targets in 2022: A Guide to Design*, International Renewable Energy Agency, Abu Dhabi, https://mc-cd8320d4-36a1-40ac-83cc-3389-cdn-endpoint.azureedge.net/-/media/Files/IRENA/Agency/Publication/2022/Nov/IRENA_RE_targets_2022.pdf?rev=f39ae339801e4853a2a0ebdb4d167f83.

IRENA (2022c), *Renewable Power Generation Costs in 2021*, International Renewable Energy Agency, Abu Dhabi, https://irena.org/publications/2022/Jul/Renewable-Power-Generation-Costs-in-2021.

IRENA (2022d), *Renewable Energy Statistics*, International Renewable Energy Agency, Abu Dhabi, www.irena.org/Data/Downloads/IRENASTAT (accessed 3 November 2022).

IRENA (2022e), *Bioenergy for the Energy Transition: Ensuring Sustainability and Overcoming Barriers*, International Renewable Energy Agency, Abu Dhabi, www.irena.org/publications/2022/Aug/Bioenergy-for-the-Transition.

IRENA (2022f), *Off-grid Renewable Energy Statistics 2022*, International Renewable Energy Agency, Abu Dhabi, www.irena.org/Publications/2022/Dec/Off-grid-renewable-energy-statistics-2022 (accessed 1 February 2023).

IRENA (2022g), *RE-Organising Power Systems for the Transition*, International Renewable Energy Agency, Abu Dhabi, www.irena.org/Publications/2022/Jun/RE-organising-Power-Systems-for-the-Transition.

IRENA (2022h), "ETAF", https://www.irena.org/Energy-Transition/Partnerships/ETAF (accessed 13 February 2023).

IRENA (2023a), *Global Landscape of Renewable Energy Finance 2023*, International Renewable Energy Agency, Madrid, Spain.

IRENA (2023b), "Renewable Energy Benefits", International Renewable Energy Agency, Abu Dhabi, https://www.irena.org/Publications/2023/Mar/World-Energy-Transitions-Outlook-2023 (accessed 1 May 2023).

IRENA (2023c), *The Changing Role of Hydropower: Challenges and Opportunities*, International Renewable Energy Agency, Abu Dhabi.

IRENA (forthcoming), *Structuring Public Financing for Universal Energy Access*, International Renewable Energy Agency, Abu Dhabi.

Jaghory, D. (2022), "Chinese Cleantech: 2022 Marks Year of Transition for Wind and Solar Policy", www.globalxetfs.com/chinese-cleantech-2022-marks-year-of-transition-for-wind-and-solar-policy/#:~:text=The%20FIT%20for%20offshore%20wind,be%20a%20year%20of%20transition.

Kene-Okafor, T. (2021), "Zola Electric Closes $90M Funding Round to Scale Technology and Enter New Markets", *TechCrunch*, https://techcrunch.com/2021/09/23/zola-electric-closes-90m-funding-round-to-scale-technology-and-enter-new-markets/ (accessed 9 September 2022).

Lorimer, M. (2021), "Vietnam's Draft Master Plan VIII – What It Means for Renewable Energy", www.wfw.com/articles/vietnams-draft-master-plan-viii-what-it-means-for-renewable-energy/.

Malhotra, K. (2022), "Pacific Island Countries Must Transition to Renewables", *Blog, Convergence News*, www.convergence.finance/news-and-events/news/2CW2dyxSl7GjwAu6I2SJeI/view (accessed 26 January 2023).

Masterson, V. (2023), "Barbados Calls for Finance Reform to Fight Climate Change", *World Economic Forum*, www.weforum.org/agenda/2023/01/barbados-bridgetown-initiative-climate-change/ (accessed 13 February 2023).

McNally, P. (2022), "An Efficient Energy Transition: Lessons from the UK's Offshore Wind Rollout", *Tony Blair Institute for Global Change*, https://institute.global/policy/efficient-energy-transition-lessons-uks-offshore-wind-rollout.

Mutambatsere, E., and M. de Vautibault (2022), "Blended Finance Can Catalyse Renewable Energy Investments in Low-Income Countries", *World Bank blogs*, https://blogs.worldbank.org/ppps/blended-finance-can-catalyze-renewable-energy-investments-low-income-countries.

NatWest (2022), "Demand Soars for CADES's Fifth Social Bond in 2022", www.natwest.com/corporates/about-us/case-studies/demand-soars-cades-fifth-social-bond.html.

Naudé, L. (2022), "Just Energy Transition Partnership Offers Should Come as Grants, Not Loans", www.wwf.org.za/?41686/Just-Energy-Transition-Partnership-offers-should-come-as-grants-not-loans (accessed 13 February 2023).

Niseiy, S. P. (2021), "Australia Funds Renewable Energy to Electrify Rural Cambodia", *Cambodianess*, https://cambodianess.com/article/australia-funds-renewable-energy-to-electrify-rural-cambodia (accessed 3 February 2023).

Norfund (2021), "USD 127 Million Social Impact Securitization Vehicle for Off-Grid Solar Sector in Kenya", www.norfund.no/app/uploads/2021/01/Press-Release-BLK1_FINAL-21-Jan-2021-004.pdf.

OECD and IEA (2022), "Support for Fossil Fuels Almost Doubled in 2021, Slowing Progress toward International Climate Goals, according to New Analysis from OECD and IEA", *Organization for Economic Co-operation and Development and International Energy Agency*, www.oecd.org/newsroom/support-for-fossil-fuels-almost-doubled-in-2021-slowing-progress-toward-international-climate-goals-according-to-new-analysis-from-oecd-and-iea.htm.

PR Newswire (2021), "U.S. Development Finance Corporation Invests $10 Million in Nithio FI to Scale Clean Energy Financing in Africa", www.prnewswire.com/news-releases/us-development-finance-corporation-invests-10-million-in-nithio-fi-to-scale-clean-energy-financing-in-africa-301381267.html (accessed 28 November 2022).

PV Magazine (2023), "IRA to Drive $114 Billion in U.S. Renewable Energy Investments by 2031, Report Says" (2023), https://pv-magazine-usa.com/2023/01/19/ira-to-drive-114-billion-in-u-s-renewable-energy-investments-by-2031-report-says/ (accessed 2 February 2023).

PV Magazine (2021), "Oorja Secures $1 Million in Seed Funding", www.pv-magazine-india.com/2021/10/12/oorja-secures-1-million-in-seed-funding/?utm_source=dlvr.it&utm_medium=linkedin (accessed 28 November 2022).

REN21 (2022), "Renewables 2022: Global Status Report", Renewable Energy Policy Network for the 21st Century, Paris, www.ren21.net/wp-content/uploads/2019/05/GSR2022_Full_Report.pdf.

Renewable Energy Agency, Abu Dhabi (n.d.), www.irena.org/publications/2022/Jan/Renewable-Energy-Market-Analysis-Africa.

RISE (n.d.), "Methodology", Regulatory Indicators for Sustainable Energy, https://rise.esmap.org/scoring-system (accessed 9 September 2022).

The Rockefeller Foundation (2022), "New analysis demonstrates how Covid crisis and Russia's invasion of Ukraine are derailing energy development in emerging and frontier economies", www.rockefellerfoundation.org/news/new-analysis-demonstrates-how-covid-crisis-and-russias-invasion-of-ukraine-are-derailing-energy-development-in-emerging-and-frontier-economies/.

SEforAll (2021), "Energizing Finance: Understanding the Landscape 2021", www.seforall.org/publications/energizing-finance-understanding-the-landscape-2021.

SIFMA (2021), "2021 Capital Markets Fact Book, Securities Industry and Financial Markets Association", www.sifma.org/resources/research/fact-book/.

SNV (2021), "BRILHO – Boosting Companies, and Lighting Lives!", SNV Netherlands Development Organization, https://snv.org/update/brilho-boosting-companies-and-lighting-lives (accessed 28 November 2022).

Solar Thermal World (2017), *Study of the Solar Water Heating Industry in Kenya*, EED Advisory Limited, Nairobi, www.solarthermalworld.org/sites/default/files/news/file/2018-10-31/study_of_the_swh_industry_-_kenya_high_res_final.pdf.

Statista (2022), "Total Assets of Insurance Companies Worldwide from 2002 to 2022", www.statista.com/statistics/421217/assets-of-global-insurance-companies/

Sward, J. (2022), "World Bank and IMF Influence Casts Shadow over South Africa's Just Energy Transition Partnership", *Bretton Woods Project*, www.brettonwoodsproject.org/2022/12/world-bank-and-imf-influence-casts-shadow-over-south-africas-just-energy-transition-partnership/ (accessed 13 February 2023).

SWFI (2022), "Top 100 Largest Sovereign Wealth Fund Rankings by Total Assets", *Sovereign Wealth Fund Institute*, www.swfinstitute.org/fund-rankings/sovereign-wealth-fund.

Taylor, M. (2022), "Falling Costs Drive Strong Demand for Australia's Residential Solar PV", *International Renewable Energy Agency*, www.irena.org/news/expertinsights/2022/May/Falling-costs-drive-strong-demand-for-Australias-residential-solar-PV (accessed 31 January 2023).

Thinking Ahead Institute (2022), "Global Top 300 Pension Funds, September 2022", *Willis Tower Watson*, www.thinkingaheadinstitute.org/content/uploads/2022/09/PI-300-2022.pdf.

Thinking Ahead Institute (2019), "Global Pension Assets Study – 2019", *Willis Tower Watson*, www.thinkingaheadinstitute.org/research-papers/global-pension-assets-study-2019/.

UN (2021), *Theme Report on Energy Access: Towards the Achievement of SDG 7 and Net-Zero Emissions*, United Nations, New York, www.un.org/sites/un2.un.org/files/2021-twg_1-061921.pdf.

UNDP (2022), *Linking Global Finance to Small-Scale Clean Energy; Financial Aggregation for Distributed Renewable Energy in Developing Countries*, New York, www.uncclearn.org/wp-content/uploads/library/Linking-Global-Finance-to-Small-Scale-Clean-Energy.pdf.

US Congress (2022), Inflation Reduction Act of (2022), www.congress.gov/bill/117th-congress/house-bill/5376/text?q=%7B%22search%22%3A%5B%22inflation+reduction+act%22%2C%22inflation%22%2C%22reduction%22%2C%22act%22%5D%7D&r=1&s=1.

USDFC (2022), "DFC Invests $40 Million in Energy Entrepreneurs Growth Fund, Increasing Access to Off-Grid Energy in Africa", US International Development Finance Corporation, www.dfc.gov/media/press-releases/dfc-invests-40-million-energy-entrepreneurs-growth-fund-increasing-access-grid (accessed 12 January 2023).

USDFC (n.d.), "Information Summary for the Public", US International Development Finance Corporation, www.dfc.gov/sites/default/files/media/documents/9000052718.PDF.

Weiss, W. and M. Spörk-Dür (2022), *Solar Heat Worldwide: Edition 2022, Solar Heating & Cooling Programme*, International Energy Agency, Paris, www.iea-shc.org/Data/Sites/1/publications/Solar-Heat-Worldwide-2022.pdf.

Wood Mackenzie (2019), "Strategic Investments in Off-Grid Energy Access", p. 43, Wood Mackenzie, Edinburgh, Scotland.

Wood Mackenzie (2022a), "Off-Grid Renewable Investment database", https://datahub.woodmac.com/app/main#/dashboards/5d3a1511d249d18c0f001758. (Subscription required).

Wood Mackenzie (2022b), "How the Russia/Ukraine War Changes Energy Markets", www.woodmac.com/news/the-edge/how-the-russiaukraine-war-changes-energy-markets/.

World Bank (2020), "West Africa Regional Energy Trade Development Policy Financing Program", World Bank, www.worldbank.org/en/news/loans-credits/2020/07/28/west-africa-regional-energy-trade-development-policy-financing-program (accessed 13 February 2023).

World Bank (2022), "World Bank Development Indicators: Access to Electricity (% of Population)", https://data.worldbank.org/indicator/EG.ELC.ACCS.ZS (accessed 3 November 2022).

World Bank (n.d.), "World Bank Development Indicators: Population, Total—Australia, China, United States", https://data.worldbank.org/indicator/SP.POP.TOTL?locations=AU-CN-US (accessed 31 January 2023).

WWF, SACAN and IEJ (2022), "IEJ Climate Finance at COP27: A First Take On South Africa's JET-IP", World Wide Fund for Nature, South African Climate Action Network and the Institute for Economic Justice, www.iej.org.za/wp-content/uploads/2022/11/IEJ_COP-27-JET-IP-Factsheet_november-2022.pdf.

3 Application of the DEJI Systems Model

INTRODUCTION TO THE DEJI SYSTEMS MODEL

The pursuit of innovative ideas, actions, and processes is essential for how we respond to the challenges of climate change. Thus, we use the framework of innovation to introduce the DEJI Systems Model in this chapter. The theme and focus of innovation run throughout this chapter. With its capstone of "systems integration," the DEJI Systems Model is applicable to both the qualitative and the quantitative aspects of climate change and the future of energy.

"A new way of doing things" is one of the definitions of innovation. Sustainability requires new ways of doing things. Whether it is this definition or any other definition, sustainability innovation must have a buy-in from stakeholders and it must be integrated into the operating environment. This chapter presents the application of the trademarked DEJI Systems Model® to the management of sustainability innovation processes. The model provides a generic pathway for sustainability innovation design, evaluation, justification, and integration. The DEJI Systems Model (Badiru, 2012, 2019) is a good tool for ensuring that the proposed innovation fits the operating environment of the organization, the community, or the nation. The DEJI Systems Model is applicable for innovation design, evaluation, justification, and integration. Figure 3.1 illustrates the overall framework of DEJI Systems Model. More details on the model can be obtained from www.DEJImodel.com.

WHY THE DEJI SYSTEMS MODEL?

Driven by analytical tools, technical professionals tend to want to jump to the coordination and implementation stage of a project. That is, jumping to the design functionality stage while dispensing with intermediate steps, where non-technical and "soft" issues might exist. However, those intermediate stages are often more critical for systems success rather than the pure analytical foundation. Items such as needs analysis, gap analysis, user involvement, communication, cooperation, resource requirement analysis, budget flow, leadership support, workforce acceptance, project desirability, and so on are essential before getting to the end point of the project. This is the essential narrative that highlights the efficacy of the approach of DEJI Systems Model, which takes a project sequentially through the stages of design, evaluation, justification, and, finally, integration. These step-by-step stages allow important considerations, technical or otherwise, to be addressed in the project. It is of utmost importance to understand how the proposed product of the project will integrate with and align with the existing organizational framework. DEJI model

DOI: 10.1201/9781003279051-3

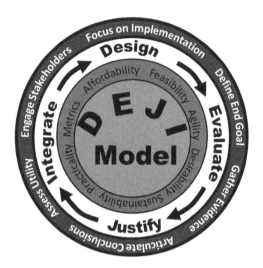

FIGURE 3.1   Elements of the DEJI Systems Model for innovation.

makes it imperative to do an a priori evaluation of the potential impact that the project output might have on the prevailing environment.

In implementation, the model can be customized for specific needs related to sustainability. Explanations and examples for design, evaluation, justification, and integration are provided throughout the chapter.

Several factors related to sustainability innovation are amenable to the application of the DEJI Systems Model. Some of these are discussed in the sections that follow. Wherever innovation is mentioned in this chapter, the specific focus is on sustainability innovation.

SUSTAINABILITY INNOVATION QUALITY MANAGEMENT

Sustainable "sustainability" requires innovation, which can come in various modes and flavors, technical or non-technical, each requiring a measure of quality management. Quality is a measure of customer satisfaction and a product's "fit-for-use" status. To perform its intended functions, a product must provide a balanced level of satisfaction to both the producer and the customer. For the purpose of sustainability pursuits, we present the following comprehensive definition of sustainability quality:

> Quality refers to an equilibrium level of functionality possessed by a product or service based on the producer's capability and the customer's needs.

Based on the above definition, quality refers to the combination of characteristics of a product, process, or service that determines the product's ability to satisfy specific needs. Quality is a product's ability to conform to specifications, where specifications represent the customer's needs or government regulations. The attainment of quality in a product is the responsibility of every employee in an organization, and the production and preservation of quality should be a commitment that extends all the way from the producer to the customer. Products that are designed to have high quality cannot maintain the inherent quality at the user's end of the spectrum if they are not used properly.

The functional usage of a product should match the functional specifications for the product within the prevailing usage environment. The ultimate judge for the quality of a product, however, is the perception of the user, and differing circumstances may alter that perception. A product that is perceived as being of high quality for one purpose at a given time may not be seen as having acceptable quality for another purpose in another time frame. Industrial quality standards provide a common basis for global commerce. Customer satisfaction or production efficiency cannot be achieved without product standards. Regulatory, consensus, and contractual requirements should be taken into account when developing product standards driven by innovation. These are described below.

Regulatory Standards

This refers to standards that are imposed by a governing body, such as a government agency. All firms within the jurisdiction of the agency are required to comply with the prevailing regulatory standards.

Consensus Standards

This refers to a general and mutual agreement between companies to abide by a set of self-imposed standards.

Contractual Standards

Contractual standards are imposed by the customer based on case-by-case or order-by-order needs. Most international standards will fall into the category of consensus standards, simply because a lack of an international agreement often leads to trade barriers.

INDUSTRY AND LEGAL STANDARDS

Along with the aforementioned standards, there are also self-coordinated standards such as industry standards and legal standards.

INNOVATIVE PRODUCT DESIGN

The initial step in any manufacturing effort is the development of a manufacturable and marketable product. An analysis of what is required for a design and what is available for the design should be conducted in the planning phase of a design project. The development process must cover analyses of the product configuration, the raw materials required, production costs, and potential profits. Design engineers must select appropriate materials, the product must be expected to operate efficiently for a reasonable length of time (reliability and durability), and it must be possible to manufacture the product at a competitive cost. The design process will be influenced by the required labor skills, production technology, and raw materials. Product planning is substantially influenced by the level of customer sophistication, enhanced technology, and competition pressures. These are all project-related issues that can be enhanced by project management. The designer must recognize changes in all these factors and incorporate them into the design process. Design project management provides a guideline for the initiation, implementation, and termination of a design effort. It sets guidelines for specific design objectives, structure, tasks, milestones, personnel, cost, equipment, performance, and problem resolutions. The steps involved include planning, organizing, scheduling, and control. The availability of technical expertise within an organization and outside of it should be reviewed. The primary question of whether or not a design is needed at all should be addressed. The "make" or "buy," "lease" or "rent," and "do nothing" alternatives to a proposed design should be among the considerations.

In the initial stage of design planning, the internal and external factors that may influence the design should be determined and given relative weights according to priority. Examples of such influential factors include organizational goals, labor situations, market profile, expected return on design investment, technical manpower availability, time constraints, state of the technology, and design liabilities. The desired components of a design plan include summary of the design plan, design objectives, design approach, implementation requirements, design schedule, required resources, available resources, design performance measures, and contingency plans.

DESIGN FEASIBILITY

The feasibility of a proposed design can be ascertained in terms of technical factors, economic factors, or both. A feasibility study is documented with a report showing all the ramifications of the design. A report of the design's feasibility should cover statements about the need, the design process, the cost feasibility, and the design effectiveness. The need for a design may originate from within the organization, from another organization, from the public, or from the customer. Pertinent questions for design feasibility review include: Is the need significant enough to warrant the proposed design? Will the need still exist by the time the design is finished? What are alternate means of satisfying the need? What technical interfaces are required for the design? What is the economic impact of the need? What is the return, financially, on the design change?

A Design Breakdown Structure (DBS) is a flowchart of design tasks required to accomplish design objectives. Tasks that are contained in the DBS collectively describe the overall design. The tasks may involve hardware products, software products, services, and information. The DBS helps to describe the link between the end objective and its components. It shows design elements in the conceptual framework for the purposes of planning and control. The objective of developing a DBS is to study the elemental components of a design project in detail, thus permitting a "divide and conquer" approach. Overall design planning and control can be significantly improved by using DBS. A large design may be decomposed into smaller sub-designs, which may, in turn, be decomposed into task groups.

Definable sub-goals of a design problem may be used to determine appropriate points at which to decompose the design.

Individual components in a DBS are referred to as *DBS elements* and the hierarchy of each is designated by a level identifier. Elements at the same level of subdivision are said to be of the same DBS level. Descending levels provide increasingly detailed definition of design tasks. The complexity of a design and the degree of control desired are used to determine the number of levels to have in a DBS. Level I of a DBS contains only the final design purpose. This item should be identifiable directly as an organizational goal. Level II contains the major subsections of the design. These subsections are usually identified by their contiguous location or by their related purpose. Level III contains definable components of the Level II subsections. Subsequent levels are constructed in more specific details depending on the level of control desired. If a complete DBS becomes too crowded, separate DBSs may be drawn for the Level II components, for example. A specification of design (SOD) should accompany the DBS. A statement of design is a narrative of the design to be generated. It should include the objectives of the design, its nature, the resource requirements, and a tentative schedule. Each DBS element is assigned a code (usually numeric) that is used for the element's identification throughout the design life cycle.

DESIGN STAGES

The guidelines for the various stages in the life cycle of a design can be summarized in the following way:

1. **Definition of design problem:** Define problem and specify the importance of the problem, emphasize the need for a focused design problem, identify designers willing to contribute expertise to the design process, and disseminate the design plan.
2. **Personnel assignment:** The design group and the respective tasks should be announced and a design manager should be appointed to oversee the design effort.
3. **Design initiation:** Arrange organizational meeting, discuss general approach to the design problem, announce specific design plan, and arrange for the use of required hardware and tools.
4. **Design prototype:** Develop a prototype design, test an initial implementation, and learn more about the design problem from test results.
5. **Full design development:** Expand the prototype design and incorporate user requirements.
6. **Design verification:** Get designers and potential users involved, ensure that the design performs as designed, and modify the design as needed.
7. **Design validation:** Ensure that the design yields the expected outputs. Validation can address design performance level, deviation from expected outputs, and the effectiveness of the solution to the problem.
8. **Design integration:** Implement the full design, ensure the design is compatible with existing designs and manufacturing processes, and arrange for design transfer to other processes.

9. **Design feedback analysis:** What are the key lessons from the design effort? Were enough resources assigned? Was the design completed on-time? Why or Why not?
10. **Design maintenance:** Arrange for continuing technical support of the design and update design as new information or technology becomes available.
11. **Design documentation:** Prepare full documentation of the design and document the administrative process used in generating the design.

CULTURAL AND SOCIAL COMPATIBILITY ISSUES

Cultural infeasibility is one of the major impediments to outsourcing sustainability innovation in a wide-open market. The business climate can be very volatile. This volatility, coupled with cultural limitations, creates problematic operational, particularly in an emerging technology. The pervasiveness of online transactions overwhelms the strict cultural norms in many markets. The cultural feasibility of information-based outsourcing needs to be evaluated from the standpoint of where information originates, where it is intended to go, and who comes into contact with the information. For example, the revelation of personal information is frowned upon in many developing countries, where there may be an interest in outsourced innovation engagements. Consequently, this impedes the collection, storage, and distribution of workforce information that may be vital to the success of outsourcing. For outsourcing to be successfully implemented in such settings, assurances must be incorporated into the hardware and software implementations so as to meet the needs of the workforce. Accidental or deliberate mismanagement of information is a more worrisome aspect of IT than it is in the Western world, where enhanced techniques are available to correct information errors. What is socially acceptable in the outsourcing culture may not be acceptable in the receiving culture and vice versa.

ADMINISTRATIVE COMPATIBILITY

Administrative or managerial feasibility involves the ability to create and sustain an infrastructure to support an operational goal. Should such an infrastructure not be in existence or unstable, then we have a case of administrative infeasibility. In developing countries, a lack of trained manpower precludes a stable infrastructure for some types of industrial outsourcing. Even where trained individuals are available, the lack of coordination makes it almost impossible to achieve a collective and dependable workforce. Systems that are designed abroad for implementation in a different setting frequently get bogged down when imported into a developing environment that is not conducive for such systems. Differences in the perception of ethics are also an issue of concern in an outsource location. A lack of administrative vision and limited managerial capabilities limit the ability of outsource managers in developing countries. Both the physical and conceptual limitations on technical staff lead to administrative infeasibility that must be reckoned with. Overzealous entrepreneurs are apt to jump on opportunities to outsource production without a proper assessment of the capabilities of the receiving organization. Most often than not outsourcing organizations don't fully understand the local limitations. Some organizations take the risk of learning as they go, without adequate prior preparation.

TECHNICAL COMPATIBILITY

Hardware maintenance and software upgrade are, perhaps, the two most noticeable aspects of technical infeasibility of information technology in a developing country. The mistake is often made that once you install IT and all its initial components, you have the system for life. This is very far from the truth. The lack of proximity to the source of hardware and software enhancement makes this situation particularly distressing in a developing country. The technical capability of the personnel as well as the technical status of the hardware must be assessed in view of the local needs. Doing an over-kill on the infusion of IT just for the sake of keeping up is as detrimental as doing nothing at all.

WORKFORCE INTEGRATION STRATEGIES

Any outsourcing enterprise requires adapting from one form of culture to another. The implementation of a new technology to replace an existing (or a nonexistent) technology can be approached through one of several cultural adaptation options. Below are some suggestions:

Parallel Interface: The host culture and the guest culture operate concurrently (side by side); with mutual respect on either side.

Adaptation Interface: This is the case where either the host culture or the guest culture makes conscious effort to adapt to each other's ways. The adaptation often leads to new (but not necessarily enhanced) ways of thinking and acting.

Superimposition Interface: The host culture is replaced (annihilated or relegated) by the guest culture. This implies cultural imposition on local practices and customs. Cultural incompatibility, for the purpose of business goals, is one reason to adopt this type of interface.

Phased Interface: Modules of the guest culture are gradually introduced to the host culture over a period of time.

Segregated Interface: The host and guest cultures are separated both conceptually and geographically. This used to work well in colonial days. But it has become more difficult with modern flexibility of movement and communication facilities.

Pilot Interface: The guest culture is fully implemented on a pilot basis in a selected cultural setting in the host country. If the pilot implementation works with good results, it is then used to leverage further introduction to other localities.

HYBRIDIZATION OF INNOVATION CULTURES

The increased interface of cultures through industrial outsourcing is gradually leading to the emergence of hybrid cultures in many developing countries. A hybrid culture derives its influences from diverse factors, where there are differences in how the local population views education, professional loyalty, social alliances, leisure pursuits, and information management. A hybrid culture is, consequently, not fully embraced by either side of the cultural divide. This creates a big challenge to managing outsourcing projects.

Sustainability quality is at the intersection of efficiency, effectiveness, and productivity. Efficiency provides the framework for quality in terms of resources and inputs required to achieve the desired level of quality. Effectiveness comes into play with respect to the application of product quality to meet specific needs and requirements of an organization. Productivity is an essential factor in the pursuit of quality as it relates to the throughput of a production system. To achieve the desired levels of quality, efficiency, effectiveness, and productivity, a new research framework must be adopted for sustainability to take hold across national and cultural boundaries.

Several aspects of quality must undergo rigorous research along the realms of both quantitative and qualitative characteristics. Many times, quality is taken for granted and the flaws only come out during the implementation stage, which may be too late to rectify. The growing trend in product recalls is a symptom of a priori analysis of the sources and implications of quality at the product conception stage. This column advocates the use of the DEJI Model for enhancing quality design, quality evaluation, quality justification, and quality integration through hierarchical and stage-by-stage processes.

Better quality is achievable and there is always room for improvement in the quality of products and services. But we must commit more efforts to the research at the outset of the product development cycle. Even the human elements of the perception of quality can benefit from more directed research from a social and behavioral sciences point of view.

SUSTAINABILITY ACCOUNTABILITY

Throughout history, engineering has answered the calls of society to address specific challenges. With such answers comes a greater expectation of professional accountability. Consider the level of social responsibility that existed during the time of the Code of Hammurabi. Two of the laws are echoed below:

Hammurabi's Law 229:

If a builder build a house for someone, and does not construct it properly, and the house which he built fall in and kill its owner, then that builder shall be put to death.

Hammurabi's Law 230:

If it kills the son of the owner the son of that builder shall be put to death.

These are drastic measures designed to curb professional dereliction of duty and enforce social responsibility with particular focus on product quality. Research and education must play bigger and more direct roles in the design, practice, and management of quality and present modern aspects of social responsibility in the context of day-to-day personal and professional activities. The global responsibility of the greater society is essential with respect to world development challenges covering the global economy, human development, global governance, and social relationships. Quality is the common theme in the development challenges. Focusing on the emerging field of Big Data, we should advocate engineering education collaboration, which aligns well with data-intensive product development. With the above principles as possible tenets for better research, education, and practice of quality in engineering and technology, this chapter suggests the DEJI model as a potential methodology. The model encourages the practice of building sustainability into a product right from the beginning so that the product integration stage can be more successful.

The design of quality in product development should be structured to follow point-to-point transformations. A good technique to accomplish this is the use of state-space transformation, with which we can track the evolution of a product from the concept stage to a final product stage. For the purpose of product quality design, the following definitions are applicable:

Product state: A state is a set of conditions that describe the product at a specified point in time. The *state* of a product refers to a performance characteristic of the product which relates input to output such that a knowledge of the input function over time and the state of the product at time $t = t_0$ determines the expected output for $t \geq t_0$. This is particularly important for assessing where the product stands in the context of new technological developments and the prevailing operating environment.

Product state space: A product *state space* is the set of all possible states of the product lifecycle. State-space representation can solve product design problems by moving from an initial state to another state, and eventually to the desired end-goal state. The movement from state to state is achieved by means of actions. A goal is a description of an intended state that has not yet been achieved. The process of solving a product problem involves finding a sequence of actions that represents a solution path from the initial state to the goal state. A state-space model consists of state variables that describe the prevailing condition of the product. The state variables are related to inputs by mathematical relationships. Examples of potential product state variables include schedule, output quality, cost, due date, resource, resource utilization, operational efficiency, productivity throughput, and technology alignment. For a product described by a system of components, the state-space representation can use state-by-state transfer of product characteristics. Each intermediate state may represent a significant milestone in the product. Thus, a descriptive state-space model facilitates an analysis of what actions to apply in order to achieve the next desired product state. A graphical representation can be developed for a product transformation from one state to another through the application of human or machine actions. This simple representation can be expanded to cover several components within the product information framework. Hierarchical linking of product elements provides an expanded transformation structure. The product state can be expanded in accordance with implicit requirements. These requirements might include grouping of design elements, linking precedence requirements (both technical and procedural), adapting to new technology developments, following required communication links, and accomplishing reporting requirements. The actions to be taken at each state depend on the prevailing product conditions. The nature of subsequent alternate states depends on what actions are implemented. Sometimes there are multiple paths that can lead to the desired end result. At other times, there exists only one unique path to the desired objective. In conventional practice, the characteristics of the future states can only be recognized after the fact, thus, making it impossible to develop adaptive plans. In the implementation of the DEJI model, adaptive plans can be achieved because the events occurring within and outside the product state boundaries can be taken into account.

If we describe a product by P state variables s_i, then the composite state of the product at any given time can be represented by a vector $\mathbf{S}$ containing P elements. The components of the state vector could represent either quantitative or qualitative variables (e.g., cost, energy, color, time). We can visualize every state vector as a point in the state space of the product. The representation is unique since every state vector corresponds to one and only one point in the state-space. Suppose we have a set of actions (transformation agents) that we can apply to the product information so as to change it from one state to another within the project state-space. The transformation will change a state vector into another state vector. A transformation may be a change in raw material or a change in design approach. The number of transformations available for a product characteristic may be finite or unlimited. We can construct trajectories that describe the potential states of a product evolution as we apply successive transformations with respect to technology forecasts. Each transformation may be repeated as many times as needed. Given an initial state $\mathbf{S}_0$, the sequence of state vectors can be represented by successive state transitions.

EVALUATION OF SUSTAINABILITY QUALITY

A product can be evaluated on the basis of cost, quality, schedule, and meeting requirements. There are many quantitative metrics that can be used in evaluating a product at this stage. Learning curve productivity is one relevant technique that can be used because it offers an evaluation basis of a product with respect to the concept of growth and decay. The half-life extension (Badiru, 2012) of the basic learning is directly applicable because the half-life of the technologies going into a product can be considered. In today's technology-based operations, retention of learning may be threatened by fast-paced shifts in operating requirements. Thus, it is of interest to evaluate the half-life properties of new technologies as they impact the overall product quality. Information about the half-life can tell us something about the sustainability of learning-induced technology performance. This is particularly useful for designing products whose life cycles stretch into the future in a high-tech environment.

JUSTIFICATION OF SUSTAINABILITY QUALITY

We need to justify an innovation program on the basis of quantitative value assessment. The Systems Value Model (SVM) is a good quantitative technique that can be used here innovation justification on the basis of value. The model provides a heuristic decision aid for comparing project alternatives. It is presented here again for the present context. Value is represented as a deterministic vector function that indicates the value of tangible and intangible attributes that characterize the product. It can be represented as V, which is the assessed value based on the contributing attributes of the components making up the product. Examples of product attributes are quality, throughput, manufacturability, capability, modularity, reliability, interchangeability, efficiency, and cost performance. Attributes are considered to be a combined function of factors. Examples of product factors are market share, flexibility, user acceptance, capacity utilization, safety, and design functionality. Factors are themselves considered to be composed of indicators. Examples of indicators are debt ratio, acquisition volume, product responsiveness, substitutability, lead time, learning curve, and scrap volume. By combining the above definitions, a composite measure of the operational value of a product can be quantitatively assessed. In addition to the quantifiable factors, attributes, and indicators that impinge upon overall project value, the human-based subtle factors should also be included in assessing overall project value.

EARNED VALUE TECHNIQUE FOR INNOVATION

Value is synonymous with quality. Thus, the contemporary earned value technique is relevant for "earned quality" analysis. This is a good analytical technique to use for the justification stage of the DEJI model. This will impact cost, quality, and schedule elements of product development with respect to value creation. The technique involves developing important diagnostic values for each schedule activity, work package, or control element. The variables are: PV: Planned Value; EV: Earned Value; AC: Actual Cost; CV: Cost Variance; SV: Schedule Variance; EAC: Estimate at Completion; BAC: Budget at Completion; ETC: Estimate to Complete. This analogical relationship is a variable research topic for quality engineering and technology applications.

INTEGRATION OF SUSTAINABILITY QUALITY

Without being integrated, a system will be in isolation and it may be worthless. We must integrate all the elements of a system on the basis of alignment of functional goals. The overlap of systems for integration purposes can conceptually be viewed as projection integrals by considering areas bounded by the common elements of subsystems. Quantitative metrics can be applied at this stage for effective assessment of the product state. Trade-off analysis is essential in quality integration. Pertinent questions include the following:

What level of trade-offs on the level of quality are tolerable?
What is the incremental cost of higher quality?
What is the marginal value of higher quality?
What is the adverse impact of a decrease in quality?
What is the integration of quality of time?

Presented below are guidelines and important questions relevant for quality integration.

- What are the unique characteristics of each component in the integrated system?
- How do the characteristics complement one another?
- What physical interfaces exist among the components?
- What data/information interfaces exist among the components?
- What ideological differences exist among the components?
- What are the data flow requirements for the components?
- What internal and external factors are expected to influence the integrated system?
- What are the relative priorities assigned to each component of the integrated system?
- What are the strengths and weaknesses of the integrated system?
- What resources are needed to keep the integrated system operating satisfactorily?
- Which organizational unit has primary responsibility for the integrated system?

The proposed approach of the DEJI model will facilitate a better alignment of product technology with future development and needs. The stages of the model require research for each new product with respect to design, evaluation, justification, and integration. Existing analytical tools and techniques can be used at each stage of the model.

UMBRELLA THEORY FOR INNOVATION

Extensive literature review concludes that an overarching theory was lacking to guide the process of innovation. The key to a successful actualization of innovation centers on how people work and behave in team collaborations. Hence, the proposed methodology of "umbrella theory for innovation" takes into account the interplay between people, tools, and process. The Umbrella Theory for Innovation capitalizes on the trifecta of human factors, process design, and technology tool availability within the innovation environment. The theory harnesses the proven efficacies of existing tools and principles of systems engineering and management. Two specific techniques in this regard are the Triple C model and the DEJI Systems Model.

A semantic network, also called a frame network, is a knowledge base that represents semantic relationships between elements in an operational network or system. It is often used for knowledge representation purposes in software systems. In innovation, a semantic network can be used to represent the relationships among elements (people, technology, and process) in the innovation system. This representation can give a visual cue of the critical paths in the innovation network. The requirements for the success of an innovation effort include the following:

1) **Relative advantage:** This is the degree to which an innovation is perceived as better than the idea it supersedes by a particular group of users, measured in terms that matter to those users, like economic advantage, social prestige, convenience, or satisfaction. The greater the perceived relative advantage of an innovation, the more rapid its rate of adoption is likely to be. There are no absolute rules for what constitutes "relative advantage". It depends on the particular perceptions and needs of the user group.

2) **Compatibility with existing values and practices:** This is the degree to which an innovation is perceived as being consistent with the values, past experiences, and needs of potential adopters. An idea that is incompatible with their values, norms or practices will not be adopted as rapidly as an innovation that is compatible.

3) **Simplicity and ease of use:** This is the degree to which an innovation is perceived as difficult to understand and use. New ideas that are simpler to understand are adopted more rapidly than innovations that require the adopter to develop new skills and understandings.

4) **Trialability:** This is the degree to which an innovation can be experimented with on a limited basis. An innovation that is triable represents less uncertainty to the individual who is considering it.

5) **Observable results:** The easier it is for individuals to see the results of an innovation, the more likely they are to adopt it. Visible results lower uncertainty and also stimulate peer discussion of a new idea, as friends and neighbors of an adopter often request information about it.

INNOVATION READINESS MEASURE

Badiru (2019) presented the framework for an innovation assessment tool. The tool is designed to assess the readiness of an organization on the basis of desired requirements with respect to pertinent factors.

Based on the spread of innovation requirements over the relevant factors, a quantitative measure of the innovation readiness of the organization can be formulated as follows:

Assuming that each checkmark can be rated on a scale of 0 to 10, the following composite measure can be derived:

$$IR = \sum_{i=1}^{N}\sum_{j=1}^{M} r_{ij},$$

Where:
IR = Innovation Readiness of the Organization
N = Number of Requirements
M = Number of Factors
r_{ij} = Alignment Measure of Requirement i with Respect to Factor j

The above measure can be normalized on a scale of 0 to 100, on the basis of which organizations and/or units within an organization can be compared and assessed for innovation readiness. Obviously, an organization that is competent in executing and actualizing innovation will yield a higher innovation readiness measure.

RISK MANAGEMENT

Risk is an essential element in any pursuit. No risk means no accomplishment. The important thing is to manage risk constructively. Risk management is an integral part of innovation. For innovation, particularly for those dealing with new ventures, risk management can be carried out effectively by investigating and identifying the sources of risks associated with each activity. These risks can be assessed or measured in terms of likelihood and impact. Because of the exploration basis of new technology, a different and diverse set of risk concerns will be involved. So, as risks are assessed for managerial processes, technical and managerial risks must also be assessed. The major activities in innovation analysis consist of feasibility studies, design, transportation, utility, survey works, construction, permanent structure works, mechanical and electrical installations, maintenance, and so on.

DEFINITION OF RISK

Risk is often ambiguously defined as a measure of the probability, level of severity, and exposure to all hazards for a project activity. Practitioners and researchers often debate the exact definition, meaning, and implications of risk. Two alternate definitions of risk are presented below:

Risk is an uncertain event or condition that, if it occurs, has a positive or negative effect on a project objective.
Risk is an uncertain event or set of circumstances that, should it occur, will have an effect on the achievement of the project's objectives.

In this book, we present the following definition of risk management:

Risk management is the state of having a contingency ready to respond to the impact (good or bad) of occurrence of risk, such that risk mitigation or risk exploitation becomes an intrinsic part of the project plan.

For any innovation undertaking, there is always a chance that things will not turn out exactly as planned. Thus, project risk pertains to the probability of uncertainties of the technical, schedule, and cost outcomes of the project. All technology-based projects are complex and they involve risks in all the phases of the project starting from the feasibility phase to the operational phase. These risks have a direct impact on the project schedule, cost, and performance. These projects are inherently complex and volatile with many variables. A proper risk mitigation plan, if developed for identified risks, would ensure better and smoother achievement of project goals within the specified time, cost and technical requirements. Conventional project management techniques, without a risk management component, are not sufficient to ensure time, cost and quality achievement of a large-scale project, which may be mainly due to changes in scope and design, changes in government policies and regulations, changes in industry agreement, unforeseen inflation, underestimation, and improper estimation. Projects, which are exposed to such risks and uncertainty, can be effectively managed with the incorporation of risk management throughout the projects' life cycle.

SOURCES OF UNCERTAINTY

Project risks originate from the uncertainty that is present in all projects to one extent or another. A common area of uncertainty is the size of project parameters, such as time, cost, and quality with respect to the expectations of the project. For example, we may not know precisely, how much time and effort will be required to complete a particular task. Possible sources of uncertainty include the following:

Poor estimates of time and cost
Lack of a clear specification of project requirements
Ambiguous guidelines about managerial processes
Lack of knowledge of the number and types of factors influencing the project
Lack of knowledge about the interdependencies among activities in the project
Unknown events within the project environment
Variability in project design and logistics
Project scope changes
Varying direction of objectives and priorities

IMPACTS OF REGULATIONS

Risks can be mitigated, not eliminated. In fact, risk is the essence of any enterprise. In spite of government regulations designed to reduce accident risks, accidents will occasionally happen. Government regulators can work with organizations to monitor data and operations. This will only preempt a fraction of potential risks of incidents. For this reason, government agencies must work with organizations to ensure that adequate precautions are taken in all operating scenarios. Government and industry must work together in a risk-mitigation partnership, rather than in an adversarial and dictatorial relationship. There is no risk-free activity in business and industry of today. For example, many of the safety and security incidents observed over the years involved human elements – errors, incompetence, negligence, and so on. How do you prevent negligence? You can encourage non-negligent operation or incentivize perfect record, but human will still be human when bad things happen. Effective risk management requires a reliable risk analysis technique. Below is how to deal with risk management:

Avoid
Assign
Assume
Mitigate
Manage

Below is a four-step process of managing risk

STEP ONE—Identify the Risks
STEP TWO—Assess the Risks
STEP THREE—Plan Risk Mitigation
STEP FOUR—Communicate Risk

We must venture out on the risk limb in order to benefit from what the innovation offers. Many leaders profess the call of "taking risk," but guidance is often lacking on to what extent risks can be taken. A quote that typifies the benefit of taking risk is echoed below:
Consider the quote below:

Behold the lowly Turtle—he only makes progress when he sticks his neck out.
—James Conan Bryan

Let us take another look at the basic definition of risk:

Risk—"Potential Realization of an Unwanted Negative Consequence"
Reward—"Potential Realization of a Desired Positive Consequence"

A master list of risk management involves the following:

- New Technology
- Functional Complexity
- New vs. Replacement
- Leverage on Company
- Intensity of Business Need
- Interface Existing Applications
- Staff Availability
- Commitment of Team
- Team Morale
- Applications Knowledge
- Client IS Knowledge
- Technical Skills Availability
- Staff Conflicts
- Quality of Information Available
- Dependability on other projects
- Conversion Difficulty
- End-date Dictate
- Conflict Resolution Mechanism
- Continued Budget Availability
- Project Standards Used
- Large/Small Project
- Size of Team
- Geographic Dispersion

- Reliability of Personnel
- Availability of Support Organization
- Availability of Champion
- Vulnerability to Change
- Stability of Business Area
- Organizational Impact
- Tight Time Frame
- Turnover of Key People
- Change Budget Accepted
- Change Process Accepted
- Level of Client Commitment
- Client Attitude toward IS
- Readiness for Takeover
- Client Design Participation
- Client Participation in Acceptance Test
- Client Proximity to IS
- Acceptance Process

Possible risk response planning can follow the following options:

Accept—Do nothing because the cost to fix is more expensive than the expected loss
Avoid—Elect not to do part of the project associated with the risk
Contingency Planning—Frame plans to deal with risk consequence and monitor risk regularly (identify trigger points)
Mitigate—Reduce either the probability of occurrence, the loss, or both
Transfer—Outsource

The perspectives and guidance offered by the Umbrella model for innovation management can create an avenue for managing, controlling, or mitigating risk in innovation pursuits.

CONCLUSIONS

It is expected that the methodology of the Umbrella theory of innovation will inspire new research inquiries into the human factors of driving innovation in organizations. The theory has the benefit of providing coverage for all the typical nuances (qualitative and quantitative) that may be encountered in the innovation environment. Of particular importance is the consideration of the people factors of innovation. The common flawed view of innovation is that it is predicated on the acquisition of technological items. While technology may be the underpinning of a specific innovation project, more often than not, the human factors will determine the success or failure of innovation. The umbrella theory explicitly calls out the people aspects of the pursuit of innovation. The quantitative measure of innovation readiness can be adapted and expanded to fit specific research, development, and implementation themes related to the pursuit of innovation. Quality is an integrative process that must be evaluated on a stage-by-stage approach. This requires research, education, and implementation strategies that consider several pertinent factors. This column suggests the DEJI Systems Model, which has been used successfully for product development applications, as a viable methodology for quality design, quality evaluation, quality justification, and quality integration. This column is intended as a source to spark the interest of researchers to apply this tool in new product development efforts. Further details and examples of the diverse applications of the DEJI Systems Model can be found in Badiru (2012, 2014, 2015, 2019, 2023), Badiru and Maloney (2016), and Badiru et al. (2015).

REFERENCES

Badiru, Adedeji B. (2012), "Application of the DEJI Model for Aerospace Product Integration," *Journal of Aviation and Aerospace Perspectives (JAAP)*, Vol. 2, No. 2, pp. 20–34, Fall.

Badiru, A. B. (2014), "Quality Insights: The DEJI Model for Quality Design, Evaluation, Justification, and Integration," *International Journal of Quality Engineering and Technology*, Vol. 4, No. 4, pp. 369–378.

Badiru, Adedeji B. (2015), "A Systems Model for Global Engineering Education: The 15 Grand Challenges," *Engineering Education Letters*, Vol. 1, No. 1, pp. 1–14. Open Access Link: 2015:3 https://doi.org/10.5339/eel.2015.3.

Badiru, Adedeji B. (2019), *Systems Engineering Models: Theory, Methods, and Applications*, Taylor & Francis Group/CRC Press, Boca Raton, FL.

Badiru, Adedeji B. (2023), *Systems Engineering Using DEJI Systems Model: Design, Evaluation, Justification, and Integration with Case Studies and Applications*, Taylor & Francis Group/CRC Press, Boca Raton, FL.

Badiru, Adedeji B. and Anna E. Maloney (2016), "A Conceptual Framework for the Application of Systems Approach to Intelligence Operations: Using HUMINT to Augment SIGINT," *American Intelligence Journal*, Vol. 33, No. 2, pp. 41–46.

Badiru, Adedeji B., John Elshaw, and Ibrahim Ade Badiru (2015), "Quality Insights: Systems-based Product Quality Assessment for Customer Preferences," *International Journal of Quality Engineering and Technology (IJQET)*, Vol. 5, No. 3/4, pp. 266–280.

4 Lean Principle and Waste Reduction

INTRODUCTION TO LEAN THINKING FOR THE ENVIRONMENT

Haste makes waste, just as rush makes ruin.

– Adedeji B. Badiru

If we apply the principles of lean operations to the problems of climate change, we can figure out how to clean up our acts with regard to the environment (Badiru and Kovach, 2012; Badiru and Agustiady, 2021). Lean principles form a key technique of the overall approach of systems engineering. If we don't want to ruin our environment, we will practice more lean initiatives in all we do, in business, industry, government, education, and the military.

Kilchiro Toyoda, founder of Toyota, remarked that "Every defect is a treasure, if the company can uncover its cause and work to prevent it across the corporation." The lesson that is conveyed by this insightful quote is that the elimination of waste in the production system can translate to a valuable treasure for the organization. Thus, the goal is to find and eliminate waste in the pursuit of whatever our goals are. This concept is directly applicable to the pursuit of sustainability.

"Lean" is a terminology that is well known and defined as an elimination of waste in operations through managerial principles. Many principles are comprised in the Lean concept, but the major thought to remember is the effective utilization of resources and time in order to achieve higher-quality products and ensure customer satisfaction. Remembering back, defects are anything that the customer is unhappy with and this is a term utilized in Six Sigma. Six Sigma identifies and eliminates these defects so that the customer is, in turn, satisfied. The customer is the number one focus; if they are unhappy, they will have no problem going elsewhere which most likely is a competition for the business. Coupling Lean and Six Sigma will reduce waste and reduce defects. In what follows this concept will be labeled Lean Six Sigma.

The most basic concept when discussing waste reduction begins with Kaizen which is a Japanese concept defined as "taking apart and making better". This concept takes a vast amount of project management techniques to facilitate the process going forward. 5s processes are the most predominant and commonly known for Kaizen events. 5s principles of this system are determined by finding a place for everything and everything in its place.

The 5s levels are as follows:

Sort—Identify and eliminate necessary items and dispose of unneeded materials that do not belong in an area. This reduces waste, creates a safer work area, opens space, and helps visualize processes. It is important to sort

DOI: 10.1201/9781003279051-4

through the entire area. The removal of items should be discussed with all personnel involved. Items that cannot be removed immediately should be tagged for subsequent removal.

Sweep—Clean the area so that it looks like new and clean it continuously. Sweeping prevents an area from getting dirty in the first place and eliminates further cleaning. A clean workplace indicates high standards of quality and good process controls. Sweeping should eliminate dirt, build pride in work areas, and build value in equipment.

Straighten—Have a place for everything and everything being in its place. Arranging all necessary items is the first step. It shows what items are required and what items are not in place. Straightening aids efficiency; items can be found more quickly and employees travel shorter distances. Items that are used together should be kept together. Labels, floor markings, signs, tape, and shadowed outlines can be used to identify materials. Shared items can be kept at a central location to eliminate purchasing more than needed.

Schedule—Assign responsibilities and due dates to actions. Scheduling guides sorting, sweeping, and straightening and prevents regression to unclean or disorganized conditions. Items are returned where they belong and routine cleaning eliminates the need for special cleaning projects. Scheduling requires checklists and schedules to maintain and improve neatness.

Sustain—Establish ways to ensure maintenance of manufacturing or process improvements. Sustaining maintains discipline. Utilizing proper processes will eventually become routine. Training is key to sustaining the effort and involvement of all parties. Management must mandate the commitment to housekeeping for this process to be successful.

The benefits of 5s include: (a) a cleaner and safer workplace; (b) customer satisfaction through better organization; and (c) increased quality, productivity, and effectiveness.

The term Kaizen is a combination of two separate terms: 'Kai' means to break apart or disassemble so that one can begin to understand; 'Zen' means to improve. This process focuses on improvements objectively by breaking down the processes in a clearly defined and understood manner so that wastes are identified, improvements are created, and wastes are both identified and eliminated. The philosophy includes reducing cycle times and lead times, in turn, increasing productivity, reducing Work-In-Process (WIP), reducing defects, increasing capacity, increasing flexibility, and improving layouts through visual management techniques.

Operator cycle times need to be understood in order to reduce the non-productive times. Operators should also be cross-functional so that they are able to perform different job functions and the workloads of each function are well balanced. The work performed needs to be not only value added work, but also work that is in demand through customers. WIP should be eliminated to reduce inventory. Inventory should be seen simply as money waiting in process and should be reduced as much as possible. WIP can be reduced by reducing set-up times, transporting smaller quantities of batch outputs, and line balancing. Bottlenecks should be removed by finding non-value-added tasks and removing the excess time spent by both machinery and humans. Flexible layouts promote efficiency by paying attention to improvement opportunities in the 5M's, as listed below:

1. Manpower
2. Machines
3. Materials
4. Methods

Sometimes an 8th dimension of Waste is added in and an abbreviation of DOWNTIME is associated with the acronym. It is defined below:

* Defect/correction
* Overproduction
* Waiting
* Not utilizing employee talents
* Transportation/material movement
* Inventory
* Motion
* Excessive processing

REDUCING WASTE AND DEFECTS

The primary technique for reducing waste and defects utilizing Lean Six Sigma techniques is demonstrated below in order of operations:
    Root cause analysis by defining the problem

1) Process mapping
2) Data gathering—Gather data on any process with defects and issues. Utilize Voice of the Customer (VOC) to find what data is needed
3) Cause/effect analysis (seeking root cause) utilizing 5M's
4) Verifying root cause with data-driven results—ask why 5 times to ensure the proper root cause is found. Do not band-aid problems, instead eliminate the cause they are occurring.
5) Solutions and continuous improvement plans
6) Test implementation plan by piloting. Pilot plans can include actual trials or mock trials with details information.
7) Implement continuous improvement ideas
8) Control/monitoring plan
9) Documentation of lessons learned

Now that the process is laid out in terms of making proper improvements, the sustainability portion must be realized. The control plan during this phase is crucial. The control plan must not just be documented, but should be a living document that is followed structurally. The accountability portion for this phase is a key portion in having lasting results. The more specific the plan is, the better off the implementation of it will be. The plans must also be attainable, or the plan will fail.

Lean can also involve some statistical tools. The tools demonstrate the efficiencies and labor balancing. The main statistical tools are as follows:

First Pass Yield (FPY) indicates the number of good outputs from a first pass at a process or step. The formula is as follows:

$$FPY = (\# \text{ accepted}) / (\# \text{ processed})$$

The formula for the First Pass Yield ratio is % FPY = [(# accepted)/(# processed)] × 100.

This number does not include re-worked product that was previously rejected.

Rolled throughput yield (RTY) covers an entire process. If a process involves three activities with FPYs of 0.90, 0.94, and 0.97, the RTY would be 0.90 × 0.94 × 0.97 = 0.82. The %RTY = 0.82 × 100 = 82%.

Value-Added Time (VAT) is % VAT = (sum of activity times)/(lead time) × 100. When the sum of activity times equals lead time, the value-added time is 100%. For most processes, % VAT = 5 to 25%. If the sum of activity times equals the lead time, the time value is not acceptable and activity times should be reduced.

Takt time is a kaizen tool used in the order taking phase. *Takt* is a German word for pace and is defined as time per unit. This is the operational measurement to keep production on track. To calculate Takt time, the formula is time available/production required. Thus, if a required production is 100 units per day and 240 minutes are available, the Takt time = 240/100 or 2.4 minutes to keep the process on track. Individual cycle times should be balanced to the pace of Takt time. To determine the number of employees required, the formula is (labor time/unit)/Takt time. Takt, in this case, is time per unit. Takt requires visual controls and helps reduce accidents and injuries in the workplace. Monitoring inventory and production WIP will reduce waste or muda. Muda is a Japanese term for waste where waste is defined as any activity that consumes some type of resource, but is non-value added for the customer. The customer is not willing to pay for this resource because it is of no benefit to them. Types of muda include scrap, re-work, defects, mistakes, and excess transport, handling, or movement.

The Lean house is a common methodology for understanding lean and waste reduction.

Mistake Proofing is a subject of its own when brought into the Lean Six Sigma methodology. This term is often called Poka Yoke, also known as another initiative to improve production systems. The methodology eliminates product defects before they occur by simply installing processes to prevent the mistakes from happening in the first place. These mistakes that happen are due to human nature and cannot normally be eliminated by simple training or SOP's. These steps to eliminate the defect will prevent the next step in the process from occurring if a defect is found. Normally, there is some type of alert that will show there is a mistake and will fail the process from going forward. An example of a Poka Yoke would be a simple check weigher that would kick off a package of food if it were not the correct weight.

Poka Yokes often also encompass a concept called Zero Quality Control (ZQC). This does not mean a reduction in defects, but instead the complete elimination of defects, also known as zero defects. ZQC was another concept led by the Japanese that leads to low-inventory production. The reason the inventory is so low is because there is no need for the firm to hold excess inventory due to having to replace less defective parts as often. ZQC also focuses on quality control and data rather than blaming humans on mistakes. The methodology was developed by Shigeo Shingo, who knew it was human nature to make common mistakes and did not feel people should be reprimanded for them. In their words, "Punishment makes people feel bad, it does not eliminate defects."

This concept is important because it focuses on the customers and realizes that defects are costly, and that therefore eliminating defects saves money. Many companies "re-work" product to save money, but do not realize the need to eliminate the problem in the first place. This process will eliminate re-work by eliminating any defects from happening in the first place.

The first cycle of the ZQC system is the Plan-Do-Check cycle also known as PDC. This is a traditional cycle where processes and conditions are planned out, the planned actions are performed in the Do phase, and finally quality-control checks are performed in the Check phase. This method catches mistakes and also provides feedback during the Check phase. The checks in this place also account for 100% inspection, and therefore all parts or processes are looked upon as indicating no defects.

There are three main types of checks or inspections that are popular:

• Judgment inspections
• Informative inspections
• Source inspections

Judgment inspections are those that are done normally by humans based on their expectations. They find the defect after it has already occurred. Informative Inspections are based on statistical quality control (SQC), checks on each product, and self-checks. These inspections help reduce defects, but do not eliminate them completely. Finally, the source inspections are those inspections that reduce the defects completely. Source inspections discover the mistakes before processing and then provide feedback and corrective actions so that the process has zero defects. The source inspections require 100% inspection. The feedback loop is also very quick so that there is minimal waiting time.

How to Use Poka Yokes

Poka Yokes use two different approaches:

• Control systems
• Warning systems

Control systems stop the equipment when a defect or un-expected event occurs. This prevents the next step in the process from occurring so that the complete process is not performed. Warning systems signal operators to stop the process or address the issue at the time. Obviously the first of the two prevents all defects and has a more ZQC methodology because an operator could be distracted or not have time to address the problem. Control systems often also use lights or sounds to bring attention to the problem that way the feedback loop again is very minimal.

The conclusion of Poka Yokes is to use the methodology as mistake-proofing for ZQC to eliminate all defects, not just some. The types of Poka Yokes do not have to be complex or expensive, they simply have to well thought out to prevent human mistakes or accidents.

The Poka Yoke discussion stems into the correct location discussion. This technique places design and production operations in the correct order to satisfy customer demand. The concept is to increase throughput of machines, ensuring the production is performed at the proper time and place. Centralization of areas helps final assemblers, but the most common practice to be effective is to unearth an effective flow. U-shaped flows normally prevent bottlenecks. Value stream mapping is a key component during this time in order to establish all steps occurring are adding value. A reminder for value-added activities: this can be any activity that the customer is willing to pay for. Another note to remember is to not only have a smart and efficient technique, but also only produce goods that the customer is demanding to eliminate excess inventory.

This technique is called the pull technique, which is the practice of not producing any goods upstream if the downstream customer does not need it. The reason this is a difficult technique is because once an efficient method is found to produce a good, the mass production begins. The operations forget if the goods are actually needed or not and begin thinking only of throughput. Even though co-manufacturers seem like a bad idea for many employers, they sometimes come in handy when needed a small amount of a versatile product.

Push systems, on the other hand, are not effective due to predictions of customer demands.

Lean systems show the pull system utilizing machinery for 90% of requirements and limits downtime to 10% for changeovers and maintenance. This does not mean preventative maintenance should not be performed, but only that the maintenance time is reduced to 10%. Kanbans are a key factor during this lean system, in order to use a visual indicator that another part or process is required. This also prevents excess parts from being made or excess processes being performed.

Heinjunka is the leveling of production and scheduling based on volume and product mix. Instead of building products according to the flow of customer orders, this technique levels the total volume of orders over a specific time so that uniform batches of different product mixes are made daily. The result is a predictable matrix of product types and volumes. For heinjunka to succeed, changeovers must be managed easily. They must be as minimally invasive as possible to prevent time wasted because of product mix. Another key to heinjunka is making sure that products are needed by customers. A product should not be included in a mix simply to produce inventory if it is not demanded by customers. Long changeovers should be investigated to determine the reason and devise a method to shorten them.

CONCLUSIONS

The pursuit of lean concepts in sustainability can be effective for success. The recommended approach is to implement lean in a step-by-step approach, as summarized in the five steps below:

1) Getting started—plan out the appropriate steps. This will take one to six months.
2) Create the new organization and restructure. This will take six to twenty-four months.
3) Implement lean techniques and systems and continually improve. This will take two to four years.
4) Complete the transformation. This will take up to five years.
5) Do the entire process again to have another continuous improvement project and sustain the results.

Readers interested in further expositions of the contents of this chapter are referred to Badiru and Kovach (2012), Agustiady and Badiru (2013), Badiru and Agustiady (2021), and George et al. (2004) for additional references and ideas on Lean thinking for operational excellence. When Lean thinking is combined with Six Sigma (see next chapter), we can have a powerful integrated methodology to address global environmental challenges.

REFERENCES

Agustiady, Tina and Adedeji B. Badiru (2013), *Sustainability: Utilizing Lean Six Sigma Techniques*, Taylor & Francis Group/CRC Press, Boca Raton, FL.

Badiru, Adedeji B. and Tina Agustiady (2021), *Sustainability: A Systems Engineering Approach to the Global Grand Challenge*, Taylor & Francis Group/CRC Press, Boca Raton, FL.

Badiru, Adedeji B. and Tina Kovach (2012), *Statistical Techniques for Project Control*, Taylor & Francis Group/CRC Press, Boca Raton, FL.

George, Mike, Dave Rowlands, and Bill Kastle (2004), *What is Lean Six Sigma?*, McGraw-Hill, New York, NY.

5 Six Sigma Methodology for Climate Response

INTRODUCTION TO SIX SIGMA FOR THE ENVIRONMENT

A bad system will defeat a good person every time.

– W. Edwards Deming

Regarding environmental sustainability and climate change challenges, we have been told again and again that pursuing efficiency, consistency, and effectiveness in our activities (private, public, or governmental) can go a long way in contributing to the remediation of the degradation of the environment. Six Sigma methodology focuses on achieving efficiency, consistency, improvement, effectiveness, and productivity in all sectors. It is one thing to reduce waste; it is another thing to minimize the chances of waste in the first place. This is benefit of Six Sigma.

The functions of quality control and improvement play a significant role in all sectors of the economy (George et al., 2004). In the present global connectivity, the traditional geographical boundaries imposed on business and industry continue to disappear (Badiru and Agustiady 2021). Businesses, manufacturers, service providers and their associated supply-chain providers are mutually entangled in a globally-competitive environment. Survival and advancement of organizations are contingent upon meeting and exceeding customer requirements. In the view of this chapter, one primary customer is the environment. Addressing our environment needs can be satisfied through an emphasis on continuous improvement of quality, which can contribute to achieving UN's sustainability development goals, as they affect the environment.

Six Sigma is about achieving uniformity in actions and products. Although it is normally applied in production settings, it has been found to be equally applicable and effective in other general applications, particularly those related to environmental issues. If we can use Six Sigma to make our production systems more efficient, more effective, and less wasteful, we can move more rapidly towards accomplishing the 17 Sustainable Development Goals (SDGs) of the United Nations.

LEAN AND GREEN PRODUCTION

In clean and green production lies our potential for a sweeping success. The goal is to make a larger proportion of our industries operating in ways that support the environment. For example, we can move industry toward low emissions from energy use in selected sectors, including the following:

- Food and agriculture
- Buildings and smart cities
- Oil & gas industry

DOI: 10.1201/9781003279051-5

- Chemical industry
- Cement industry
- Iron and steel industry
- Information and communications technologies

The Six Sigma methodology can have an impact in all industries. Sustainability activities, such as recycling, reusing, recovering, or redesigning, can benefit from the business techniques of Six Sigma. Since this is a book on climate change and the environment, the full details of the techniques of Six Sigma are not provided. This chapter is designed to present a brief overview, for which further exposition can be explored in more comprehensive references. For details and examples of many of the techniques in this chapter, readers may refer to Agustiady and Badiru (2013).

Six Sigma is best defined as a business process improvement approach that seeks to find and eliminate causes of defects and errors, reduce cycle times, reduce costs of operations, improve productivity, meet customer expectations, achieve higher asset utilization, and improve return on investment (ROI). Six Sigma deals with producing data-driven results through management support of the initiatives (Agustiady and Badiru, 2013).

Six Sigma pertains to sustainability because without the actual data, decisions would be made on trial and error. Sustainable environments require having actual data to back up decisions so that methods are used to have improvements for future generations. The basic methodology of Six Sigma includes a five-step method approach that consists of the following:

Define: Initiate the project, describe the specific problem, identify the project's goals and scope, and define key customers and their Critical to Quality (CTQ) attributes.

Measure: Understand the data and processes with a view to specifications needed for meeting customer requirements, develop and evaluate measurement systems, and measure current process performance.

Analyze: Identify potential cause of problems, analyze current processes, identify relationships between inputs, processes, and outputs, and carry out data analysis.

Improve: Generate solutions based on root causes and data-driven analysis while implementing effective measures.

Control: Finalize control systems and verify long-term capabilities for sustainable and long-term success.

The goal for Six Sigma is to strive for perfection by reducing variation and meeting customer demands. The customer is known to make specifications for processes. Statistically speaking, Six Sigma is a process that produces 3.4 defects per million opportunities. A defect is defined as any event that is outside of the customer's specifications. The opportunities are considered any of the total number of chances for a defect to occur.

SIX-SIGMA SPREAD

The Greek letter σ (sigma) marks the distance on the horizontal axis between the mean μ and the curve inflection point. The greater the distance, the greater is the

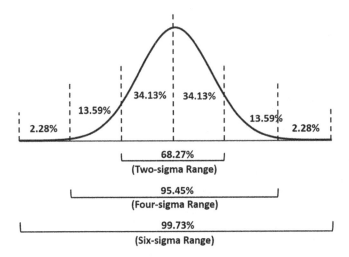

FIGURE 5.1 Areas under the normal curve for Six Sigma.

spread of values encountered. The genesis of Six Sigma is represented by the curve in Figure 5.1. The figure shows a mean of 0 and a standard deviation of 1, that is, $\mu = 0$ and $\sigma = 1$. The plot also illustrates the areas under the normal curve within different ranges around the mean. The upper and lower specification limits (USL and LSL) are $\pm 3 \sigma$ from the mean or within a Six Sigma spread. Because of the properties of the normal distribution, values lying as far away as $\pm 6 \sigma$ from the mean are rare because most data points (99.73%) are within $\pm 3 \sigma$ from the mean, except for processes that are seriously out of control.

Six Sigma allows no more than 3.4 defects per million parts manufactured, or 3.4 errors per million activities in a service operation. To appreciate the effect of Six Sigma, consider a process that is 99% perfect (10,000 defects per million parts). Six Sigma requires the process to be 99.99966% perfect to produce only 3.4 defects per million, that is $3.4/1,000,000 = 0.0000034 = 0.00034\%$. That means that the area under the normal curve within $\pm 6 \sigma$ is 99.99966% with a defect area of 0.00034%.

TOOLS OF SIX SIGMA

The following tools are the most common Six Sigma tools and will be explained how they are to be used in the concept of sustainability.

- Project charter
- SIPOC (Suppliers, Inputs, Process, Outputs, and Customers)
- Kano model
- CTQ (Critical to Quality)
- Affinity diagram
- Measurement system analysis
- Gage R and R
- Variation

- Graphical analysis
- Location and spread
- Process capabilities
- Cause and effect diagram
- FMEA (Failure Modes and Effects Analysis)
- Process mapping
- Hypothesis testing
- ANOVA (Analysis of Variance)
- Correlation
- Linear regression
- Theory of constraints
- SMED (Single-Minute Exchange of Dies)
- TPM (Total Productive Maintenance)
- Design for Six Sigma
- Quality Function Deployment
- DOE (Design of Experiments)
- Control charts
- Control plan

PROJECT CHARTER

A project charter is a definition of the project that includes the following:

- Provides problem statement
- Overview of scope, participants, goals, and requirements
- Provides authorization of a new project
- Identifies roles and responsibilities

Once the project charter is approved, it should not be changed.

A project charter begins with the project name, the department of focus, the focus area, and the product or process.

A project charter serves as the focus point throughout the project to ensure the project is on track and the proper people are participating and being held accountable.

The importance of a project charter in aspect to sustainability is the living document to educate and give governance for a new project. Sustainability needs to utilize a great deal of education while giving goals and objectives. A project charter will serve as this living document for organizations with specified approaches.

SIPOC

The SIPOC identifies

1. Major tasks and activities
2. The boundaries of the process
3. The process outputs
4. Who receives the outputs (the customers)
5. What the customer requires of the outputs
6. The process inputs

7. Who supplies the inputs (suppliers)
8. What the process requires of the inputs
9. The best metrics to measure

Supplier—Know and work with your supplier while making your supplier improve
Input—Strive to continually improve the inputs by trying to do the right thing the first time
Process—Describe the process at a high level, but with enough detail to demonstrate to an executive or manager. Understand the process fully by knowing it 100%. Eliminate any mistakes by doing a Poka Yoke
Output—Strive to continually improve the outputs by utilizing metrics
Customer—Keep the customer's requirements in sight by remembering they are the most important aspect of the project. The customer makes the specifications, keep the CTQ's of the customer in mind

SIPOC STEPS

1. Gain top-level view of the process
2. Identify the process in simple terms
3. Identify External Inputs such as raw materials, employees, etc.
4. Identify the Customer Requirements also known as Outputs
5. Make sure to include all value-added and non-value-added steps
6. Include both process and product output variables

SIPOC implies that the process is understood and helps easily identify opportunities for improvement.

A SIPOC is important in concepts of sustainability because it helps develops a solution for development. Normally, the process is mapped out in a well-defined, but high-level first.

The following is an example of a SIPOC for the steps to recycling a product:

Recycling Steps

1) Collect and process goods into a container that is friendly to deposit/refund or pick up programs
2) Undergo a recycling loop where products are sorted by type and parts
3) Parts go to a smelting plant where parts are put back to original state
4) Melted parts are sent back to a factory to make new products
5) Pieces are put back together in a completed product form at a factory
6) New products are sold
7) Supplier provides new product

The SIPOC map is continued with the outputs:

Product readied for recycling
Sorted product
Melted product
Recycled product ready for production
New product

The same process occurs with the inputs:

Newspapers
Plastic
Cardboard
Glass
Magazines
Cans
Metal

The suppliers are identified:

ABC Local newspaper
Coca Cola Bottles
Box for ABC Macaroni container
Glass jar for ABC sauce
ABC Magazine
ABC canned soup

Finally, the customer is realized. It is important to understand the customer is not always the end customer and is also part of the recycling process.

Trash recycling pick-up
Container for recycling
Recycling factory
Sorting personnel
Melting personnel
Completed smelting process organizer
Delivery of recycled product to factory
Factory of re-production
Customer buying product

The important part of a SIPOC is to look at the details of the current state and see what improvements can be made for future states. Adding specifications for any of the inputs can identify gaps in the process. Benchmarking one process to another will also identify gaps.

KANO MODEL

The Kano model was developed by Noriaki Kano in the 1980s. The Kano model is a graphical tool that further categorizes VOC & CTQs into three distinct groups:

- Must Haves
- Performance
- Delighters

The Kano helps identify CTQ's that add incremental value versus those that are simply requirements where having more is not necessarily better. The Kano model engages customers by understanding the product attributes which are most likely important to customers. The purpose of the tool is to support product specifications which are made by the customer and promote discussion while engaging team members. The model differentiates features of products rather than customer needs by understanding necessities and items that are not required whatsoever. Kano also produced a methodology for mapping consumer responses with questionnaires that focused on attractive qualities through reverse qualities. The five categories for customer preferences are as follows:

- Attractive
- One-dimensional
- Must-be
- Indifferent
- Reverse

Attractive qualities are those which provide satisfaction when fulfilled; however, they do not result in dissatisfaction if not fulfilled.

One-dimensional qualities are those which provide satisfaction when fulfilled, and dissatisfaction if not fulfilled.

Must-be qualities are those which are taken for granted if fulfilled, but provide dissatisfaction when not fulfilled.

Indifferent qualities are those which are neither good nor bad, resulting in neither customer satisfaction nor dissatisfaction.

Reverse qualities are those which result in high levels of dissatisfaction from some customers and show that most customers are not alike.

The Kano model is important to use when being sustainable because it is important to differentiate which aspects we must accomplish to protect our environment and which aspects we can gradually improve upon.

CTQ

CTQ (Critical To Quality) represent the key measurable characteristics of a product or process whose performance standards or specification limits must be met in order to satisfy the customer. These outputs represent the product or service characteristics defined by the customer (internal or external). CTQs are important to the customer. They come from the VOC. CTQs are measurable and quantifiable metrics that come from the VOC.

An affinity diagram is an organizational tool for articulating the VOCs.

Elements that are Critical to Quality are critical to sustainability because we need to understand the critical aspects to the environment that matter most. The following are the main sources of resource consumption by rank:

- Electricity—#1 Source of Resource Consumption
- Natural Gas—#2 Source of Resource Consumption
- Water/Sewer—#3 Source of Resource Consumption

Therefore, the CTQ characteristics are electricity, natural gas, and water/sewer.

Utilizing a VOC for manufacturing internally is a good way to understand processes that the employees know a great deal about. Therefore, the production worker(s) are the customers and the questions are given to them. Based on the two questions asked and the responses, the machinery is the CTQ attribute since it affects more than one of the problems.

AFFINITY DIAGRAM

An affinity diagram is a tool conducted to place large amounts of information into an organized manner by grouping the data into characteristics. The steps for an affinity diagram are as follows:

- Step 1: Clearly define the question or focus of the exercise
- Step 2: Record all participant responses on note cards or post-it notes
- Step 3: Lay out all note cards or post the post-its onto a wall
- Step 4: Look for and identify general themes
- Step 5: Begin moving the note cards or post-it notes into the themes until all responses are allocated
- Step 6: Re-evaluate and make adjustments

The exact same methodology for a basic process can be done for a manufacturing or business process where sustainability is in question. The pros and cons are then sought after to reach a decision. The decision should be made by having a consensus from the group where the pros outweigh the cons.

MEASUREMENT SYSTEMS ANALYSIS

Gage R&R (gage repeatability and reproducibility)

Gage R&R is a measurement systems analysis (MSA) technique that uses continuous data based on the principles that:

- Data must be in statistical control
- Variability must be small compared to product specifications
- Discrimination should be about one-tenth of product specifications or process variations
- Possible sources of process variation are revealed by measurement systems
- Repeatability and reproducibility are primary contributors to measurement errors
- The total variation is equal to the real product variation plus the variation due to the measurement system
- The measurement system variation is equal to the variation due to repeatability plus the variation due to reproducibility
- Total (observed) variability is an additive of product (actual) variability and measurement variability

Discrimination is the number of decimal places that can be measured by the system. Increments of measure should be about one-tenth of the width of a product specification or process variation that provides distinct categories.

Accuracy is the average quality near to the true value.

The *true value* is the theoretically correct value.

Bias is the distance between the average value of the measurement and the true value, the amount by which the measurement instrument is consistently off-target, or systematic error. *Instrument accuracy* is the difference between the observed average value of measurements and the true value. Bias can be measured based by instruments or operators. Operator bias occurs when different operators calculate different detectable averages for the same measure. Instrument bias results when different instruments calculate different detectable averages for the same measure.

Precision encompasses total variation in a measurement system, the measure of natural variation of repeated measurements, and repeatability and reproducibility.

Repeatability is the inherent variability of a measurement device. It occurs when repeated measurements are made of the same variable under absolutely identical condition (same operators, set-ups, test units, environmental conditions) in the short term. Repeatability is estimated by the pooled standard deviation of the distribution of repeated measurements and is always less than the total variation of the system.

Reproducibility is the variation that results when measurements are made under different conditions. The different conditions may be operators, setups, test units, or environmental conditions in the long term. Reproducibility is estimated by the standard deviation of the average of measurements from different measurement conditions.

The *measurement capability index* is also known as the precision-to-tolerance (P/T) ratio. The equation is P/T = (5.15 × σMS)/tolerance. The P/T ratio is usually expressed as a percent and indicates what percent of the tolerance is taken up by the measurement error. It considers both repeatability and the reproducibility. The ideal ratio is 8% or less; an acceptable ratio is 30% or less. The 5.15 standard deviation accounts for 99% of MS variation and is an industry standard.

The P/T ratio is the most common estimate of measurement system precision. It is useful for determining how well a measurement system can perform with respect to the specifications. The specifications, however, may be inaccurate or need adjustment. The %R&R = (σMS/σTotal) × 100 formula addresses the percent of the total variation taken up by measurement error and includes both repeatability and reproducibility.

A Gage R&R can also be performed for discrete data also known as binary data. This data is also known as yes/no or defective/non-defective-type data. The data still requires at least 30 data points. The percentages of repeatability, reproducibility, and compliance should be measured. If no repeatability is able to be shown, there will also be no reproducibility. The matches should be above 90% for the evaluations. A good measurement system will have a 100% match for repeatability, reproducibility, and compliance.

If the result is below 90%, the operational definition must be re-visited and re-defined. Coaching, teaching, mentoring, and standard operating procedures should be reviewed, and the noise should be eliminated. A decision needs to be made on which activities and production equipment are sustainable.

The Gage R&R bars are desired to be as small as possible, driving the Part-to-Part bars to be larger.

The averages of each operator is different, meaning the reproducibility is suspect. The operator is having a problem making consistent measurements.

The Operator*Samples interactions lines should be reasonably parallel to each other. The operators are not consistent to each other.

The measurement by samples graph shows there is minimal spread for each sample and a small amount of shifting between samples.

The Sample*Operator of .706 shows that the interaction was not significant which is what is wanted from this study.

The % contribution part to part of 10.81 shows the parts are the same.

The total Gage R&R % study variation of 94.44, % contribution of 89.19, tolerance of 143.25, and distinct categories of 1 showed that there was no repeatability, reproducibility, and was not a good Gage. The number of categories being less than 2 shows the measurement system is of minimal value since it will be difficult to distinguish one part from another.

The Gage Run Chart shows that there is no consistency between measurements.

Conclusion, B1 is the best Blender, and Dominic is the best operator. There is no reproducibility or repeatability between any of the measurements.

VARIATION

Variation is present in all processes, but the goal is to reduce the variation while understanding the root cause of where the variation comes from in the process, and why. For Six Sigma to be successful, the processes must be in control statistically and the processes must be improved by reducing the variation. The distribution of the measurements should be analyzed to find the variation and depict the outliers or patterns.

The study of variation began with Dr. W. E. Deming, who was also known as the Father of Statistics. Deming stated that variation happens naturally, but the purpose is to utilize statistics to show patterns and types of variations. There are two types of variations that are sought after, special cause variation and common cause variation. Special cause variation refers to out-of-the-ordinary events such as a power outage, whereas common cause variation is inherent in all processes and is typical. The variation is sought to be reduced so that the processes are predictable, in statistical control, and have a known process capability. A root cause analysis should be done on special cause variation so that the occurrence is not to happen again. Management is in charge of common cause variation where action plans are given to reduce the variation.

Assessing the location and spread are important factors as well. Location is known as the process being centered along with the process requirements. Spread is known as the observed values compared to the specifications. The stability of the process is required. The process is said to be in statistical control if the distribution of the measurements have the same shape, location, and spread over time. This is the point in time where all special causes of variation are removed and only common cause variation is present.

An *average, central tendency* of a dataset is a measure of the "middle" or "expected" value of the dataset. Many different descriptive statistics can be chosen as measurements of the central tendency of the data items. These include the arithmetic mean, the median, and the mode. Other statistical measures such as the standard deviation and the range are called measures of spread of data. An average is a single value meant to represent a list of values. The most common measure is the arithmetic mean but there are many other measures of central tendency, such as the median (used most often when the distribution of the values is skewed by small numbers with very high values).

As stated before, special cause variation would be occurrences such as power outages, large mechanical breakdowns, etc. Common cause variations would be occurrences such as electricity being different by a few thousand kilowatts per month. In order to understand the variation, graphical analyses should be done followed by capability analyses.

It is important to understand the variation in the systems so that the best-performing equipment is used. The variation sought after is, in turn, utilized for sustainability studies. The best-performing equipment should be utilized most and the least-performing equipment should be brought back to its original state of condition and then upgraded or fixed to be capable. Capability indices are explained next.

PROCESS CAPABILITIES

The capability of a process is the spread that contains most of the values of the process distribution. Capability can only be established on a process that is stable with a distribution that only has common cause variation.

CAPABLE PROCESS (C_p)

A process is capable ($C_p \geq 1$) if its natural tolerance lies within the engineering tolerance or specifications. The measure of process capability of a stable process is $6\hat{\sigma}$, where $\hat{\sigma}$ is the inherent process variability that is estimated from the process. A minimum value of $C_p = 1.33$ is generally used for an ongoing process. This ensures a very low reject rate of 0.007% and therefore is an effective strategy for the prevention of nonconforming items. C_p is defined mathematically as

$$C_p = \frac{\text{USL} - \text{LSL}}{6\sigma}$$
$$= \frac{\text{allowable process spread}}{\text{actual process spread}}$$

where
  USL = upper specification limit
  LSL = lower specification limit

C_p measures the effect of the inherent variability only. The analyst should use R-bar/ d_2 to estimate $\hat{\sigma}$ from an R-chart that is in a state of statistical control, where R-bar is the average of the subgroup ranges and d_2 is a normalizing factor that is tabulated for different subgroup sizes (n). We don't have to verify control before performing a capability study. We can perform the study, then verify control after the study with the use of control charts. If the process is in control during the study, then our estimates of capabilities are correct and valid. However, if the process was not in control, we would have gained useful information, as well as proper insights as to the corrective actions to pursue.

CAPABILITY INDEX (C_{PK})

Process centering can be assessed when a two-sided specification is available. If the capability index (C_{pk}) is equal to or greater than 1.33, then the process may be adequately centered. C_{pk} can also be employed when there is only a one-sided specification. For a two-sided specification, it can be mathematically defined as:

$$C_{pk} = \text{Minimum} \left\{ \frac{\text{USL} - \bar{X}}{3\hat{\sigma}}, \quad \frac{\bar{X} - \text{LSL}}{3\hat{\sigma}} \right\}$$

where
$\bar{X}$ = Overall process average

However, for a one-sided specification, the actual C_{pk} obtained is reported. This can be used to determine the percentage of observations out of specification. The overall long-term objective is to make C_p and C_{pk} as large as possible by continuously improving or reducing process variability, $\hat{\sigma}$, for every iteration so that a greater percentage of the product is near the key quality characteristics target value. The ideal is to center the process with zero variability.

If a process is centered but not capable, one or several courses of action may be necessary. One of the actions may be that of integrating a designed experiment to gain additional knowledge on the process and in designing control strategies. If excessive variability is demonstrated, one may conduct a nested design with the objective of estimating the various sources of variability. These sources of variability can then be evaluated to determine what strategies to use in order to reduce or permanently eliminate them. Another action may be that of changing the specifications or continuing production and then sorting the items. Three characteristics of a process can be observed with respect to capability, as summarized below:

1. The process may be centered and capable.
2. The process may be capable but not centered.
3. The process may be centered but not capable.

PARETO CHART AND GRAPHICAL ANALYSIS

Graphical analyses are visual representations of tools that show meaningful key aspects of projects. These tools are commonly known as dotplots, histograms, normality plots, Pareto diagrams, second-level Pareto's (also known as stratification), boxplots, scatter plots, and marginal plots. The plotting of data is a key beginning step to any type of data analysis because it is a visual representation of the data. If a particular manufacturing company wants to understand where majority of their electrical costs are coming from while trying to reduce those costs, a Pareto chart and other graphical analyzes can be very useful. A Pareto chart is a powerful graphical tool for separating the "important few" from the "trivial many." To address a complex problem, we would construct Pareto charts at many levels to dig deeper down into the problem. Further details on using a Pareto diagram can be found in Badiru and Kovach (2012).

PROCESS MAPPING

The importance of process mapping is to depict all functions in the process flow while understanding if the functions are value- or non-value-added. Any delays are to be eliminated and decisions are meant to be as efficient as possible. The purpose of process mapping is to have a visual image of the process.

CAUSE AND EFFECT DIAGRAM

After a process is mapped, the cause and effect (C&E) diagram can be completed. This process is so important because it completed root cause analysis. The basis behind root cause analysis is to ask "Why?" five times in order to get to the actual root cause. Many times problems are "band aided" in order to fix the top-level problem, but the actual problem itself is not addressed.

The fish bone is broken out to the most important categories in an environment:

- Measurements
- Material
- Personnel
- Environment

METHODS

- Machines

This process requires a team to do a great deal of brainstorming where they focus on the causes of the problems based on the categories. The "fish head" is the problem statement.

FAILURE MODE AND EFFECT ANALYSIS (FMEA)

In order to select action items from the C and E diagram and prioritize the projects, FMEAs are completed. The FMEA will identify the causes, assess risks, and determine further steps. The steps to an FMEA are the following:

1. Define process steps.
2. Define functions.

3. Define potential failure modes.
4. Define potential effects of failure.
5. Define the severity of a failure.
6. Define the potential mechanisms of failure.
7. Define current process controls.
8. Define the occurrence of failure.
9. Define current process control detection mechanisms.
10. Define the ease of detecting a failure.
11. Multiply severity, occurrence, and detection to calculate a risk priority number (RPN).
12. Define recommended actions.
13. Assign actions with key target dates to responsible personnel.
14. Revisit the process after actions have been taken to improve it.
15. Recalculate RPNs with the improvements.

What can be seen from the FMEA is that an important aspect to sustainability is the RPN number reducing after the action items. It is important to understand the processes severity to a customer and increasing the capability of the process to, in turn, improve the process. The RPN's reduction will make the entire process more sustainable by being able to deliver the process at the best capabilities through thorough project management. It is important to maintain the FMEA so that once a process is improved it is not forgotten about.

HYPOTHESIS TESTING

Hypothesis testing validates assumptions made by verification of the processes based on statistical measures. It is important to use at least 30 data points for hypothesis testing so that there is enough data to validate the results.

Normality of the data points must be found in order for the hypothesis testing to be accurate.

The assumptions are shown in the null and alternate hypothesis:

Ho = (The null hypothesis): The difference is equal to the chosen reference value $\mu_1 - \mu_2 = 0$

Ha = (The alternate hypothesis): The difference is not equal to the chosen reference value $\mu_1 - \mu_2$ is not = 0

95% CI for mean difference: (1.16, 6.69) T-Test of mean difference = 0 (vs not = 0): T-value = 2.90 P-value = 0.007

The confidence interval for the mean difference between the two materials does not include zero, which suggests a difference between them. The small p-value ($p = 0.007$) further suggests that the data are inconsistent with H0: $\mu\, d = 0$; that is, the two materials do not perform equally. Specifically, the first set (mean = 79.697) performed better than the next set (mean = 83.623) in terms of weight control over the time span. Conclusion, reject Ho, the difference is not equal to the chosen reference value: $\mu_1 - \mu_2$ is not $= 0$.

ANOVA

The purpose of an ANOVA (also known as Analysis of Variance) is to determine if there is a relationship between a discrete, independent variable and a continuous, dependent output. There is a one-way ANOVA which includes one-factorial variance and a two-way ANOVA, which includes a two-factorial variance. Three sources of variability are sought after:

Total—Total variability within all observations
Between—Variation between subgroup means
Within—Random chance variation within each subgroup, also known as noise

The equation for a one-way ANOVA is:

$$SS_T = SS_F + SS_e$$

The principles for the one-way ANOVA and two-way ANOVA are the same except that in a two-way ANOVA, the factors can take on many levels. The total variability equation for a two-way ANOVA is:

$$SS_T = SS_A + SS_B + SS_{AB} + SS_e$$

where
SS_T = total sum of squares
SS_F = sum of squares of the factor
SS_e = sum of squares from error
SS_A = sum of squares for factor A
SS_B = sum of squares for factor B
SS_{AB} = sum of squares due to interaction of factors A and B

If the ANOVA shows that at least one of the means is different, a pairwise comparison is done to show which means are different. The residuals, variance, and normality should be examined and the main effects plot and interaction plots should be generated.

The F-ratio in an ANOVA compares the denominator to the numerator to see the amount of variation expected. When the F-ratio is small, which is normally close to 1, the value of the numerator is close to the value of the denominator, and the null hypothesis cannot be rejected stating the numerator and denominator are the same. A large F-ratio indicates the numerator and denominator are different, also known as the MS error, where the null hypothesis is rejected.

Outliers should also be sought after in the ANOVA, showing the variability is affected.

The main effects plot shows the mean values for the individual factors being compared. The differences between the factor levels can be seen with the slopes in the lines. The p-values can help determine if the differences are significant. Interaction plots show the mean for different combinations of factors.

CORRELATION

The linear relationship between two continuous variables can be measured through correlation coefficients. The correlation coefficients are values between -1 and 1.

If the value is around 0, there is no linear relationship.
If the value is less than .05, there is a weak correlation.
If the value is less than .08, there is a moderate correlation.
If the value is greater than .08, there is a strong correlation.
If the value is around 1, there is a perfect correlation.

SIMPLE LINEAR REGRESSION

The regression analysis describes the relationship between a dependent and independent variable as a function $y = f(x)$.

The equation for simple linear regression as a model is:

$$Y = b_0 + b_1 x + E$$

Y is the dependent variable
b_0 is the axis intercept
b_1 is the gradient of the regression line
x is the independent variable
E is the error term or residuals

The predicted regression function is tested with the following formula:

$$R^2 = \frac{\text{SSTO} - \text{SSE}}{\text{SSTO}}$$

where

$$\text{SSTO} = \begin{cases} \underline{Y}' - n\bar{Y}^2, & \text{if constant} \\ \underline{Y}\,\underline{Y}, & \text{if not constant} \end{cases}$$

Note: When the no constant option is selected, the total sum of square is uncorrected for the mean. Thus, the R^2 value is of little use, since the sum of the residuals is not zero.

The F-test shows if the predicted model is valid for the population and not just the sample. The model is statistically significant if the predicted model is valid for the population.

The regression coefficients are tested for significance through t-tests with the following hypothesis:

H_o: $b_0 = 0$, the line intersects the origin
H_A: $b_0 \neq 0$, the line does not intersect the origin
H_o: $b_1 = 0$, there is no relationship between the independent variable xi and the
dependent variable y
H_A: $b_1 \neq 0$, there is a relationship between the independent variable xi and the
dependent variable y

After the inverse relationship is seen, a regression analysis can be performed.

An example is shown below for the analysis of whether there was a pressure degradation over time on a particular piece of equipment:

A linear relationship was sought after. First, it was sought to see if there was correlation since it can be seen that there is a linear relationship between the variables. The y was the measurement and the x was the time.

THEORY OF CONSTRAINTS

Dr. Eliyahu M. Goldratt created a theory of constraints (TOC). This management theory proved that every system has at least one constraint limiting it from 100% efficiency. The analysis of a system will show the boundaries of the system. TOC not only shows the cause of the constraints, but also provides a way to resolve the constraints. There are two underlying concepts with TOC:

1) System as chains
2) Throughput, inventory management, and operating expenses

The performance of the entire system is called the chain. The performance of the system is based on the weakest link of the chain or the constraint. The remaining links are known as non-constraints. Once the constraint is improved, the system becomes more productive or efficient, but there is always a weakest link or constraint. This process continues until there is 100% efficiency.

If there are three manufacturing lines and they produce the following:

1) 250 units/day
2) 500 units/day
3) 600 units/day

The weakest link is manufacturing line 1 because it produces the least amount of units/day. The weakest link is investigated until it reaches the capacity of the non-constraints. After the improvement has been made, the new weakest link is investigated until the full potential of the manufacturing lines can be fulfilled without exceeding market demand. If the external demand is fewer than the internal capacity, it is known as an external constraint.

THROUGHPUT ANALYSIS

Throughput can be defined as (sales price – variable cost)/time. Profits should be understood when dealing with throughput. Inventories are known as raw materials, unfinished goods, purchased parts, any investments made. Inventory should be seen as dollars on shelves. Any inventory is a waste unless utilized in a just-in-time manner.

Operating expenses should include all expenses utilized to produce a good. The less the operating expenses, the better. These costs should include direct labor, utilities, supplies, and depreciation of assets.

Applying the TOC concept helps to make the weakest link stronger. There are five steps to the process of TOC:

1) Identify the constraint or the weakest link.
2) Exploit the constraint by making it as efficient as possible without spending money on the constraint or considering upgrades.
3) Subordinate everything else to the constraint—adjust the rest of the system so the constraint operates at its maximum productivity. Evaluate the improvements to ensure the constraint has been addressed properly and it is no longer the constraint. If it is still the constraint, complete the steps, otherwise skip step 4.
4) Elevate the constraint—This step is only required if steps 2 and 3 were not successful. The organization should take any action on the constraint to eliminate the problem. This is the process where money should be spent on the constraint or upgrades should be investigated.
5) Identify the next constraint and begin the five-step process over. The constraint should be monitored and continuous improvement should be completed.

SINGLE-MINUTE EXCHANGE OF DIES (SMED)

What is SMED?
Single-Minute Exchange of Die, which consists of the following:

- Theory and set of techniques to make it possible to perform equipment setup and changeover operations in under 10 minutes
- Originally developed to improve die press and machine tool setups, but principles apply to changeovers in all processes
- It may not be possible to reach the "single-minute" range for all setups, but SMED dramatically reduces setup times in almost EVERY case
- Leads to benefits for the company by giving customers variety of products in just the quantities they need
- High quality, good price, speedy delivery, less waste, cost-effective

It is important to understand large lot production which leads to trouble.

The three key topics to consider when understanding large lot production are the following:

- Inventory waste
  - Storing what is not sold costs money
  - Ties up company resources with no value to the product
- Delay
  - Customers have to wait for the company to produce entire lots rather than just what they want
- Declining quality
  - Storing unsold inventory increases chances of product being scrapped or re-worked, adding costs

Once this is realized, the benefits of SMED can be understood:

- Flexibility
  - Meet changing customer needs without excess inventory
- quicker delivery
  - Small-lot production equals less lead time and less customer waiting time
- Better quality
  - Less inventory storage equals fewer storage-related defects
  - Reduction of setup errors and elimination of trial runs for new products
- Higher productivity
  - Reduction in downtime
  - Higher equipment productivity rate

Two types of operations are realized during setup operations, which consist of internal and external operations. Internal setup can only be done when the machine is shut down (i.e. a new die can only be attached to a press when the press is stopped).

External setup can be done while the machine is still running (i.e. bolts attached to a die can be assembled and sorted while the press is operating).

It is important to convert as much internal work as possible to external work.

Four important questions to ask yourself when understanding SMED are the following:

- How might SMED benefit your factory?
- Can you see SMED benefiting you?
- What operations are internal operations?
- What operations are external operations?

There are three stages to SMED which are defined below:

- Separate internal and external setup
  - Distinguish internal vs external
  - By preparing and transporting while the machine is running can cut changeover times by as much as 50%

- Convert internal setup to external setup
  - Re-examine operations to see whether any steps are wrongly assumed as internal steps
  - Find ways to convert these steps to external setups
- Streamline all aspects of setup operations
  - Analyze steps in detail
  - Use specific principles to shorten time needed especially for steps internally with machine stopped

Five Traditional setup steps are also defined:

- Preparation—Ensures that all the tools are working properly and are in the right location
- Mounting and extraction—Involves the removal of the tooling after the production lot is completed and the placement of the new tooling before the next production lot
- Establishing control settings—Setting all the process control settings prior to the production run. Inclusive of calibrations and measurements needed to make the machine, tooling operate effectively
- First run capability—This includes the necessary adjustments (re-calibrations, additional measurements) required after the first trial pieces are produced
- Setup improvement—The time after processing during which the tooling, machinery is cleaned, identified, and tested for functionality prior to storage

The three stages of SMED are explained next.
Description of Stage 1—separate internal vs. external setup:
Three techniques help us separate internal vs. external setup tasks:

1) Use checklists
2) Perform function checks
3) Improve transport of die and other parts

Checklists: A checklist lists everything required to set up and run the next operation. The list includes items such as

- Tools, specifications, and workers required
- Proper values for operating conditions such as temperature, pressure, etc.
- Correct measurement and dimensions required for each operation
- Checking item of the list before the m/c is stopped helps prevent mistakes that come up after internal set up begun

Function checks:

- Should be performed before setup begins so that repair can be made if something does not work right
- If broken dies, molds, or jigs are not discovered until test runs are done, a delay will occur in internal setup

- Make sure such items are in working order before they are mounted will cut down setup time a great deal

Improved transport of parts and tools:

- Dies, tools, jigs, gauges, and other items needed for an operation must be moved between storage areas and machines, then back to storage once a lot is finished
- To shorten the time the machine is shut down, transport of these items should be done during external setup
- In other words, new parts and tools should be transported to the machine before the machine is shut down for changeover

Description of Stage 2—Convert internal setups to external setups:

I. Advance preparation of conditions:
   - Get necessary parts, tools and conditions ready before internal set up begins
   - Conditions like temperature, pressure, or position of material can be prepared externally while the machine is running (i.e. pre-heating of mold/material)
II. Function standardization:
   - It would be expensive and wasteful to make external dimensions of every die, tool, or part the same, regardless of the size or shape of the product it forms. Function standardizations avoid this waste by focusing on standardizing only those elements whose functions are essential to the set up
   - Function standardization might apply to dimensioning, centering, securing, expelling, or gripping
III. Implementing function standardization with two steps:
   - Look closely at each individual function in your setup process and decide which functions, if any, can be standardized
   - Look again at the functions and think about which can be made more efficient by replacing the fewest possible parts (i.e. Clamping function standardization)

Description of Stage 3—Streamline all aspects of the setup operation:

- External setup improvement includes streamlining the storage and transport of parts and tools.
- In dealing with small tools, dies, jigs, and gauges, it is vital to address issue of tool and die management.

Ask Q's Like (Ask Questions Like):

- What is the best way to organize these items?
- How can we keep these items maintained in perfect condition and ready for the next operation?
- How many of these items should we keep in stock?

Improving storage and transport:

*   Operation for storing and transporting dies can be very time-consuming, especially when your factory keeps a large number of dies on hand
*   Storage and transport can be improved by marking the dies with color codes and location numbers of the shelves where they are stored

Streamlining internal setup:

*   Implement parallel operations, using functional clamps, eliminating adjustments, and mechanization

Implementing parallel operations:

Machines such as plastic molding machines and die-casting machines often require operation at both the front and back of the machine. One-person changeovers of such machines mean wasted time and movement because the same person is constantly walking back and forth from one end of the machine to the other.
Parallel operations divide the setup operation between two people, one at each end of the machine. When setup is done using parallel operations, it is important to maintain reliable and safe operations and minimize waiting time. To help streamline parallel operations, workers should develop and follow procedural charts for each setup.

The final understanding of SMED comes from basic principles such as observing with videos.
If there is nobody in the screen, it means there is waste present.
It is important to understand that SMED is more than just a series of techniques. It is a fundamental approach to improvement activities. A personal action plan should be found to adhere to each business's needs. It important to find ways to implement SMED into environments to continue the sustainability of the businesses. To begin the process, a communication plan should be implemented.

TPM (TOTAL PRODUCTIVE MAINTENANCE)

TPM is a well-known activity which has been given several different names associated within. Many people associate TPM with the terms total predictive maintenance or total preventative maintenance. The association explained below will be total productive maintenance, but it also includes the above.
TPM is performed in the improve phase based on downtimes or efficiency losses. The downtimes associated can be either planned or unplanned. The goal of TPM is to increase all operational equipment efficiencies to above 85% by eliminating any wasted time such as setup time (see SMED section), idle times, downtimes, start-up delays, and any quality losses.

TPM ensures minimal downtime, but, in turn, also requires no defects. There are three basic steps for TPM, each of which contain several steps.

1) Analyze the current processes
   a. Calculate any costs associated with maintenance
   b. Calculate overall equipment effectiveness (OEE) by finding the proportion of quality products produced at a given line speed
2) Restore equipment to its original and high operating states
   a. Inspect the machinery
   b. Clean the machinery
   c. Identify necessary repairs on the machinery
   d. Document defects
   e. Create a scheduling mechanism for maintenance
   f. Ensure maintenance has repaired machinery and improvements are sustained
3) Preventative maintenance to be carried out
   a. Create a schedule for maintenance with priorities—include high machinery defects, replacement parts, and any information pertaining
   b. Create stable operations—complete root cause analysis on high machinery defects and machinery that causes major downtime
   c. Create a planning and communication system—documentation of preventative maintenance activities should be accessible to all people so planning and prioritization within is completed
   d. Create processes for continuous maintenance—inspections should occur regularly and servicing for any machinery should be noted on a scheduled basis
   e. Internal operations should be optimized—any internal operations should be benchmarked with improvements from other areas to eliminate time spent on root cause analysis. When defects of machinery are not understood, it is important to put the machinery back to its original state to understand the root causes more efficiently. Time to exchange parts or retrieve parts should also be minimized.
   f. Continuous improvement on preventative maintenance—train employees for early detection of problems and maintenance measures. Visual controls should be put in place for changeovers. 5S should take place to eliminate wasted time. The documentation should be communicated and plans should be presented regularly. All aspects should be looked upon to see if continuous improvements can be made.

The key TPM indicators will be able to show the following main issues:

• OEE
• Mean time between failures
• Mean time to repair

TPM is crucial to sustainability because it involves all the employees, including high-level managers and creates planning for preventative maintenance so the issues are fixed before they become an error or defect. TPM also is a journey for educating and training the work force to be familiar of machinery, parts, processes, and damages while being productive.

DFSS (DESIGN FOR SIX SIGMA)

DFSS (Design for Six Sigma) is another process that is included in a phase called **DMADV** (Define, Measure, Analyze, Design, Verify), which is a Six Sigma framework that focuses primarily on the development of a new service, product, or process as opposed to improving a previously existing one. This DMADV approach is especially useful when implementing new strategies and initiatives because of its basis in data, its early identification of success and a thorough analysis. This could be useful in developing new environmental-response initiatives. Further, if this could be mapped to the DEJI Systems Model, a more robust response strategy could be developed.

DMAIC (DEFINE, MEASURE, ANALYZE, IMPROVE, AND CONTROL)

DMAIC represents the five phases that make up the process, consisting of defining the problem, the improvement activity, the opportunity for improvement, the project goals, and customer (internal and external) requirements. DMAIC is a data-driven quality strategy used to improve processes. It is an integral part of a Six Sigma initiative, but in general can be implemented as a standalone quality improvement procedure or as part of other process-improvement initiatives such as lean. Both DMAIC and DMADV can be embedded in the DEJI Systems Model presented earlier in Chapter 3.

DMAIC differs from DMADV in terms of the design and verification portions. DMAIC is process-improvement-driven whereas DMADV is concerned with designing new products or services. Design stands for the designing of new processes required, including their implementation. Verify stands for the results being verified and the performance of the design to be maintained. The purpose of DFSS is very similar to the regular DMAIC cycle where it is a customer-driven design of processes with Six Sigma capabilities. DFSS does not only have to be manufacturing-driven; the same methodologies can be used in service industries. The process is top-down with flow-down CTQs that match flow-up capabilities. DFSS is quality-based, with predictions being made regarding first-pass quality. The quality measurements are driven through predictability in the early design phases. Process capabilities are utilized to make final design decisions. Finally, process variances are monitored to verify if Six Sigma customer requirements are met. The main tools utilized in DFSS are FMEAs, Quality Function Deployment (QFD), Design of Experiments (DOE), and simulations.

Summary of DMAIC Steps:

1. **Define** the problem, improvement activity, opportunity for improvement, the project goals, and customer (internal and external) requirements.

- Project charter to define the focus, scope, direction, and motivation for the improvement team.
- VOC to understand feedback from current and future customers, indicating offerings that satisfy, delight, and dissatisfy them.
- Value stream map to provide an overview of an entire process, starting and finishing at the customer, and analyzing what is required to meet customer needs.

2. **Measure** process performance.
- Process map for recording the activities performed as part of a process.
- Capability analysis to assess the ability of a process to meet specifications.
- Pareto chart to analyze the frequency of problems or causes.

3. **Analyze** the process to determine root causes of variation, poor performance (defects).
- Root cause analysis (RCA) to uncover causes.
- Failure Mode and Effects Analysis (FMEA) for identifying possible product, service, and process failures.
- Multi-vari chart to detect different types of variation within a process.

   Note: the term "multi-vari chart" is a more recent term used to describe a visual way to display analysis of variance data.

4. **Improve** process performance by addressing and eliminating the root causes.
- DOE to solve problems from complex processes or systems where there are many factors influencing the outcome and where it is impossible to isolate one factor or variable from the others.
- Kaizen events to introduce rapid change by focusing on a narrow project and using the ideas and motivation of the people who do the work.

5. **Control** the improved process and future process performance.
- Control plan to document what is needed to keep an improved process at its current level.
- Statistical process control (SPC) for monitoring process behavior.
- 5S to create a workplace suited for visual control.
- Mistake-proofing (poka-yoke) to make errors impossible or immediately detectable.

QFD (QUALITY FUNCTION DEPLOYMENT)

Dr. Yoji Akao developed Quality Function Deployment (QFD) in 1966 in Japan. There was a combination of quality assurance and quality control that led to value engineering analyses. The methods for QFD are simply to utilize consumer demands into designing quality functions and methods to achieve quality into subsystems and specific elements of processes. The basis for QFD is to take customer requirements from the VOC and relaying them into engineering terms to develop products or services. Graphs and matrices are utilized for QFD. A house-type matrix is compiled to ensure the customer needs are being met into the transformation of the processes or services designed. The QFD house is a simple matrix where the legend is used to understand quality characteristics, customer requirements, and completion.

DOE (DESIGN OF EXPERIMENTS)

Design of Experiments is an experimental design that shows what is useful, what is a negative connotation, and what has no effect. The majority of the time, for example, 50% of the designs have no effect.

DOEs require a collection of data measurements, systematic manipulation of variables also known as factors placed in a pre-arranged way (experimental designs), and control for all other variables. The basis behind DOEs are to test everything in a pre-arranged combination and measure the effects of each of the interactions.

The following DOE terms are used:

- Factor: An independent variable that may affect a response
- Block: A factor used to account for variables that the experimenter wishes to avoid or separate out during analysis
- Treatment: Factor levels to be examined during experimentation
- Levels: Given treatment or setting for an input factor
- Response: The result of a single run of an experiment at a given setting (or given combination of settings when more than one factor is involved)
- Replication (replicate): Repeated run(s) of an experiment at a given setting (or given combination of settings when more than one factor is involved)

There are two types of DOEs: Full-factorial design and fractional factorial design.

Full-factorial DOEs determine the effect of the main factors and factor interactions by testing every factorial combination.

A full-factorial DOE factors all levels combined with one another covering all interactions. The effects from the full-factorial DOE can then be calculated and sorted into main effects and effects generated by interactions.

Effect = Mean value of response when factor setting is at high level (Y_A+) – mean value of response when factor setting is at low level (Y_A-)

In a full factorial experiment, all of the possible combinations of factors and levels are created and tested.

In a two-level design (where each factor has two levels) with k factors, there are 2^k possible scenarios or treatments. This can be calculated as below:

- 2 factors each with 2 levels, we have $2^2 = 4$ treatments
- 3 factors each with 2 levels, we have $2^3 = 8$ treatments
- k factors each with 2 levels, we have 2^k treatments

The analysis behind the DOE consists of the following steps:

1) Analyze the data.
2) Determine factors and interactions.
3) Remove statistically insignificant effects from the model, such as p-values of less than 1, and repeat the process.
4) Analyze residuals to ensure the model is set correctly.

5) Analyze the significant interactions and main effects on graphs while setting up a mathematical model.
6) Translate the model into common solutions and make sustainable improvements.

A fractional factorial design locates the relationship between influencing factors in a process and any resulting processes while minimizing the number of experiments. Fractional factorial DOEs reduce the number of experiments while still ensuring the information lost is as minimal as possible. These types of DOEs are used to minimize time spent, money spent, and eliminating factors that seem unimportant.
The formula for a fractional factorial DOE is:

$$2^{k-q}, \text{ where q equals the reduction factor.}$$

The fractional factorial DOE requires the same number of positive and negative signs as a full-factorial DOE.

CONTROL CHARTS

Control charts are a great interpretation of understanding whether projects are being sustained by utilizing process monitoring. The process spread can be understood through control charts while also interpreting whether the process is in control and predictable. Common cause and special cause variation is able to be found through control charts. The amount of samples taken is an important aspect to control charts along with the frequency of sampling. It is important to have a random, yet normal pattern of data. For example, if data is taken during normal operating conditions and then one data point is taken during a changeover, the data will be skewed and show a point out of control.

X-BAR AND RANGE CHARTS

The R-chart is a time plot useful for monitoring short-term process variations. The X-bar chart monitors longer-term variations where the likelihood of special causes is greater over time. Both charts utilize control lines called upper and lower control limits and central lines; both of these types of lines are calculated from process measurements. They are not specification limits or percentages of the specifications or other arbitrary lines based on experience. They represent what a process is capable of doing when only common cause variation exists, in which case the data will continue to fall in a random fashion within the control limits and the process is in a state of statistical control. However, if a special cause acts on a process, one or more data points will be outside the control limits and the process will no longer be in a state of statistical control.
The following components should be used for control chart purposes:

• **UCL**—Upper Control Limit
• **LCL**—Lower Control Limit

- **CL**—Center Line. This shows where the characteristic average falls
- **USL**—Upper specification limit or upper customer requirement
- **LSL**—Lower specification limit or lower customer requirement

Control limits describe the stability of the process. Specifically, control limits identify the expected limits of normal, random, or chance variation that is present in the process being monitored. Control limits are set by the process.

Specification limits are those limits which describe the characteristics the product or process must have in order to conform to customer requirements or to perform properly in the next operation.

Control limits describe the representative nature of a stable process. Specifically, control limits identify the expected limits of normal, random, or chance variation that is present in the process being monitored. Sustainable processes must follow these rules.

CONTROL PLANS

A control plan is a vital part of sustainability because without it there is no sustainability. A control plan takes the improvements made and ensures that they are being maintained and continuous improvement is achieved. A control plan is a very detailed document that includes who, what, where, when, and why (the why is based on the root cause analysis). The 12 basic steps of a control plan are listed below:

1) Collect existing documentation for the process
2) Determine the scope of the process for the current control plan
3) Form teams to update the control plan regularly
4) Replace short-term capability studies with long-term capability results
5) Complete control plan summaries
6) Identify missing or inadequate components or gaps
7) Review training, maintenance, and operational action plans
8) Assign tasks to team members
9) Verify compliance of actual procedures with documented procedures
10) Retrain operators
11) Collect sign-offs from all departments
12) Verify effectiveness with long-term capabilities

CONCLUSIONS

A control plan ensures consistency while eliminating as much variation from the system as possible. The plans are essential to operators because it enforces SOPs and eliminates changes in processes. It also ensures PMs are performed and the changes made to the processes are actually improving the problem that was found through the root cause analysis. Control plans hold people accountable if reviewed at least quarterly. As an example, a house is designed for operational stability and sustainability that encompass the topics discussed earlier. The house can be re-arranged or re-worded with different goals and tools set out for the particular business or manufacturing environment. The base of the house shows the stability and the areas needed for a stable and sustainable work environment. The pillars are the goals and tools utilized within the house in order for the roof to be stable upon the pillars. The roof of the house is the ultimate goal for a beneficial environment. Sustainability provides the foundation for achieving structural longevity in our global infrastructure.

REFERENCES

Agustiady, Tina and Adedeji B. Badiru (2013), *Sustainability: Utilizing Lean Six Sigma Techniques*, Taylor & Francis Group/CRC Press, Boca Raton, FL.

Badiru, Adedeji B. and Tina Agustiady (2021), *Sustainability: A Systems Engineering Approach to the Global Grand Challenge*, Taylor & Francis Group/CRC Press, Boca Raton, FL.

Badiru, Adedeji B. and Tina Kovach (2012), *Statistical Techniques for Project Control*, Taylor & Francis Group/CRC Press, Boca Raton, FL.

George, Mike, Dave Rowlands, and Bill Kastle (2004), *What is Lean Six Sigma?*, McGraw-Hill, New York, NY.

6 Legal and Business Aspects of Climate Agreements

INTRODUCTION

This chapter is based on Wagenknecht et al. (2022). It addresses the qualitative issues related to the legal and business aspects of executing climate change agreements. The original report was specifically on COP26 (Conference of the Parties, 26th edition), but the discussions are relevant for all future COP conventions.

LEGAL ASPECTS

REGULATORY IMPLEMENTATION OF RULES CONCERNING CLIMATE CHANGE

Within the framework of international climate diplomacy nations have committed themselves to achieve certain goals. The current US administration pledged to cut greenhouse gas emissions 50–52 percent below 2005 levels in 2030, reaching a 100 percent carbon pollution-free power sector by 2035, and achieving a net-zero economy by no later than 2050.

Whereas the European Union announced to cut greenhouse gas emissions by 55 percent below 1990 levels in 2030, and also achieving a net-zero economy by no later than 2050.

Every nation and supranational organization like the European Union (EU) are free to decide whether to cut greenhouse gas emissions, to which amount, and on how to deliver on these goals. This is in line with the aforementioned conclusion that there are operational differences among nations and goals need to be adapted and integrated into the prevailing local practices of the target system. Hence, it is also important to note that from a regulatory point of view the commitments from each nation are implemented by the respective nation in its sole/own discretion.

LEVERAGING LEGAL PROBLEM-SOLVING METHODOLOGY

Similar to the above-mentioned eight-step process for engineering problem solving, it is common in the legislative process to also consider potential consequences prior to passing a certain regulation.

In principle, there is prerogative in favor of the legislative branch. That means that the legislative power is in principle free to decide which measures it wants to apply in order to serve a certain matter in the public interest. Hence, measures enacted by the legislative are only subject to a limited review by the judiciary power.

DOI: 10.1201/9781003279051-6

According to European law, for example, every measure attributed to Member States and the EU itself must comply with the principle of proportionality. The principle of proportionality is used among others as a logical method to balance between the restrictions imposed by a certain measure and the intended aim. The stages of the proportionality test are as follows:

- There must be a legitimate aim for a measure
- The measure must be suitable to achieve the aim (potentially with a requirement of evidence to show it will have that effect)
- The measure must be necessary to achieve the aim, that there cannot be any less onerous way of doing it
- The measure must be reasonable, considering the competing interests of different groups at hand.

Besides, it is not only good practices to foresee any adverse effects as the case may be by way of applying the proportionality test. It has also become mandatory to conduct a so-called "regulatory impact analysis" in advance of a planned regulation.

Such approach aims to make measures by the State to become more efficient, reduce harmful effects to the absolute minimum and to consider potential alternatives. In this regard, aimed effects and unwanted disadvantages are analyzed.

In the United States Regulatory Impact Analyses are governed by guidance issued by the Office of Management and Budget (OMB), OMB Circular A-4. Circular A-4 requires agencies to clearly identify the need for regulation and its impacts.

Within the office of OMB the Office of Information and Regulatory Affairs (OIRA) reviews a Regulatory Impact Analysis for transparency, utility, and objectivity.

According to the aforementioned guideline, the regulatory impact analysis comprises the following assessment steps:

- ascertain why regulation is needed,
- consider a reasonable number of alternative regulatory approaches, and
- for each alternative conduct a rigorous and objective cost–benefit analysis.

The European Union applies a similar scheme according to the so-called "Integrated Impact Assessment." Details to such an approach are contained in different guidelines by the European Commission.

IMPLEMENTATION OF POLICIES THROUGH REGULATION

States which have pledged to decarbonize usually issue policies on how to reach the self-defined goals. In particular, the following policy tools are available in the context of decarbonization:

- Tax credits and promotion/funding of investments in new technologies, e.g.
  - Carbon capture and storage, carbon dioxide removal
  - Geoengineering the oceans
  - Solar radiation management
  - Electrification or the use of hydrogen made with renewable electricity

- Fusion energy
- Enhanced rock weathering
- Spending in climate-smart programs and infrastructure
- Sector-specific performance standards such as a clean electricity standard and zero emissions vehicle standard, e.g. clean car rules
- Carbon pricing
- Make the Federal Government lead by example by way of slashing government's greenhouse gas emissions faster and/or earlier
- Impose tariff on imports from countries that contribute to climate change
- Mandate for companies to disclose climate-related risks and opportunities

Once the policies have been agreed upon, they need to be implemented. Usually, the implementation of policies is conducted by specific regulations. Without any sufficient legal framework (and government funding/tax incentives) there would be no investments in new technologies or no actions that are supposed to support the intended decarbonization.

CARBON CAPTURE AND STORAGE, CARBON DIOXIDE REMOVAL

The carbon budget set by nations will likely be exceeded. As a consequence, according to the United Nations Intergovernmental Panel on Climate Change (IPCC) the world cannot solely rely on slashing emissions to lower the excess carbon dioxide.

However, IPCC's April 2022 report also mentions that there are strategies existing according to which excess carbon can be removed from the atmosphere. Some forest-based methods refer to planting more trees. Others deal with new technologies, e.g. carbon capture and storage. According to the IPCC, such solutions are required to decarbonize and mitigate climate change.

Carbon capture with carbon storage is the process in which carbon is caught from industrial emissions at their source, thus preventing them from entering the atmosphere at all. The carbon is then pumped underground for long-term storage.

Another similar technology is also being developed and called carbon dioxide removal. This approach aims to build huge machines to suck CO_2 out of the atmosphere and store it underground. It is also referred to as "vacuuming CO_2 from the sky".

From a legal point of view it is required to establish a legal framework for direct air capture and carbon dioxide removal in advance. Such framework needs to address the conditions for respective activities, e.g. liability of operators for certain events and a scheme to transport and store CO_2. Otherwise, there would be no certainty with regard to the specific conditions and as a consequence no respective investments in new technologies.

Since carbon capture and storage as well as carbon dioxide removal are relatively expensive technologies, substantial government funding and tax incentives are required to help demonstrate the technologies at commercial scale in different applications in the first place. Furthermore, capturing carbon adds cost and companies would understandably demand subsidies to cover the expenses. Besides, proposals to build a network of carbon dioxide pipelines might only be prepared if tax credits apply which provide companies with a certain amount of money per metric ton for capturing and storing carbon dioxide. Hence, it appears that entrepreneurial activity and corporate investment in carbon removal are at least for the time being exclusively driven by government funding and tax incentives.

Another legal issue relates to the pipelines that are required to transport the captured CO_2. There is a risk that unwilling property owners do not agree that pipelines are laid out on their property.

It might be disputed whether property rights and other concerns outweigh the potential benefits of the pipelines to local industry and the climate.

Usually, companies would conclude an agreement with landowners according to which they are entitled to install and maintain pipelines that are supposed to transport carbon captured for injection underground. If this is not feasible a seizure of land might be conducted provided that such approach is in the public interest and compensation is paid to the affected landowner. This has been practiced in the past for pipelines which were used for the transport of oil and gas.

The decisive question in this case would be whether pipelines which are supposed to transport captured CO_2 also serve a public purpose. It would then need to be assessed whether carbon capture and storage as a mean to meet the nation's climate targets qualifies as "public purpose." This question might be subject to a potential future litigation case in which a landowner challenges the seizure of his or her land.

GEOENGINEERING THE OCEANS

Carbon removal is not limited to the above-described carbon capture technologies. Carbon removal can also take place by way of making use of the oceans.

In 2021 the US National Academy of Sciences released a report which describes six different approaches on how the oceans could be used to pull even more carbon dioxide out of the atmosphere:

- Nutrient fertilization
- Artificial upwelling and downwelling
- Seaweed cultivation
- Ecosystem recovery
- Ocean alkalinity enhancement
- Electrochemical processes

However, the identified techniques are still subject to further research and are far from being implemented in the near future. The rationale for this is that any harmful (side) effects on marine life needs to be excluded. This applies for any of the aforementioned measures.

Another reason why none of the measures can be implemented directly is because of the missing legal framework. Similar to carbon capture and storage/carbon dioxide removal technologies, a legal framework is required in advance of any implementation. Even if some already existing regulations on national or international level might cover some aspects of geoengineering the oceans, there is no specific regulation regarding carbon dioxide removal by way of using the oceans.

Regulations on carbon capture and removal technologies are likely to be set up soon. This is due to the fact that these technologies are mostly considered not to have any harmful effects to the environment. Whereas regulations that govern the conditions of ocean-geoengineering technologies will most likely not be set up any time soon. It first needs to be assessed whether the specific technology in question is really effective and has no harmful effect on marine life. Accordingly, it has been acknowledged in the context of the report by the National Academy of Sciences that "all techniques have multiple hurdles to overcome, ranging from feasibility to regulatory" and moving forward will require "a deeper dive on research and regulations in tandem."

SOLAR RADIATION MANAGEMENT

The term "Solar Radiation Management" stands for different options which aim to reduce the intensity of solar radiation so that global warming will decline accordingly.

One highly disputed way of solar radiation provides for the injection of sulfur into the stratosphere. The sulfur is supposed to reflect solar radiation back into space, thus cooling down the Earth. A similar effect occurs following a major volcanic eruption. This technique is also referred to as "global dimming."

Global dimming is controversial because some fear that it might create droughts in the Amazon region as well as in regions of South Asia and West Africa. Besides, this approach would not remove any CO_2 at all.

At present, it cannot be completely ruled out that a State or a private party nonetheless starts to apply a Solar Radiation Management. This is due to the fact that specific Solar Radiation Management measures are not explicitly prohibited. Hence, some call for such explicit prohibition of any respective measures on the international level and even demand that any patent application in the context of Solar Radiation Management shall be rejected.

It can be expected that this issue will be further discussed on the international level in the framework of the United Nations.

FURTHER LEGAL ISSUES REGARDING THE IMPLEMENTATION OF SELECTED POLICIES

- **Reduction of government's greenhouse gas emissions**

As previously described, the US administration pledged to reaching a 100 percent carbon pollution-free power sector by 2035, and achieving a net-zero economy by no later than 2050.

It has to be noted that, according to a Presidential Executive Order, a special scheme applies in order to slash the US government's greenhouse gas emissions:

- By 2023: Cut greenhouse gas emissions by 65 percent
- By 2027: All "light duty" government vehicles to be electric
- By 2032: Half of all US facilities and properties to have net-zero emissions
- By 2030: Slash government's greenhouse gas emissions by 65 percent; 100 percent carbon-pollution free electricity
- By 2035: 100 percent of federally-owned vehicles to be electric; entire US electricity sector to be carbon-free

- By 2045: All 300,000 leased or owned federal properties to have net-zero emissions
- By 2050: Net-zero emissions from all federal operations including contracts for goods and services

The changes outlined in the Presidential Executive Order are designed to make the federal government lead by example.

It has to be noted that there are no rules governing the implementation of the directive yet. However, such rules are required in order to secure transparent and competitive procurement processes. For example, with regard to the future procurement of federally-owned vehicles to be electric the detailed conditions for such approach need to be further stipulated. Furthermore, following the completion of the rulemaking process the government staff has to be trained accordingly with regard to the application of the new procurement rules.

- **Mandate for companies to disclose climate-related risks and opportunities**

The US Securities and Exchange Commission (SEC) recently proposed mandatory climate risk disclosure rules for certain US companies, while in the European Union companies are addressed by the proposal for a Corporate Sustainability Reporting Directive.

According to the aforementioned rules by the SEC, such companies are obliged to share information to investors about their emissions and how they are managing risks related to climate change and future climate regulations through annual reports.

The draft rules are subject to public comment and thereafter the final rules will be issued by the SEC.

There are critics of climate disclosures which provide that the SEC has no authority to require disclosures that are not financially material. As a consequence, the disclosure requirements might be legally challenged.

The Administrative Procedure Act allows courts to vacate SEC rules that are deemed arbitrary or capricious because the agency failed to offer sufficient justification for choosing the proposal over alternatives.

However, according to the SEC the information is financially material for shareholders noting that "physical and transition risks from climate change can materialize in financial markets in the form of credit risk, market risk, insurance or hedging risk, operational risk, supply chain risk, reputational risk and liquidity risk."

If the mandatory climate risk disclosure rules will be eventually enacted by the SEC, companies might become the target of litigation. This is at least true for companies which are relevant contributors to climate, or are failing to manage the risks posed by climate change to their businesses, or are providing inadequate or incomplete disclosure.

- **Impose tariffs on imports from States that contribute to climate change**

Another policy is to impose tariffs on imports from States that contribute to climate change or on the imports of major energy-intensive goods (e.g. iron, steel, aluminum, cement, fossil fuels, and petrochemicals). Such policy does not remove or reduce any CO_2 at all. However, it (indirectly) induces other countries to make more efforts in this regard.

In general, tariffs are considered as trade barriers and are, as such, not in line with the General Agreement on Tariffs and Trade (GATT). But GATT Article XX contains a catalog of general exceptions to its obligations.

According to GATT Article XX(b), measures "necessary to protect human, animal or plant life or health" are justified. Whereas GATT Article XX(g) covers measures "relating to the conservation of exhaustible natural resources."

If a State imposes tariffs on imports from another State that contributes to climate change it might be the case that this measure will be challenged by the exporting State. It will then be decisive whether the tariffs can be regarded as justified according to GATT Article XX(b) or (g).

OVERVIEW: ECONOMIC IMPLICATIONS OF CLIMATE CHANGE

The general economic perspective of climate change-related reflections implies two main spheres. Both can be directly derived from the COP26 achievement segments as well as from the Sixth IPCC Assessment report. The two relevant spheres of economic impacts from climate change are outlined as:

Adoption and Protection
Response and Mitigation

ADOPTION AND PROTECTION SPHERE OF ECONOMIC IMPACTS

On the one hand, states, their governments, societies, and private organizations across the globe have to cope with the occurring weather and climate extremes caused by global warming. The changes in extremes, which are strengthened in their development and partially irreversible, have been observed and also physically confirmed [1], directly related to human influence on climate change. Since the magnitude of and worldwide experience with climatic changes is expected to increase as global temperatures continue to rise, adaption, protection, respectively, risk reduction measures are pervasively introduced to deal with being impacted. Following the IPCC Working Group II report contribution [2], the vulnerability to the fall-outs of climate change is substantially dependent on the region and is reinforced by several factors such as socio-economic development, governance and inequity, among others. Consequently, the adaption sphere is influenced by the regarded region and its particularities.

Supporting efforts to tackle climate impacts is as well an important part of the COP26 pledges, which include a surge in funding for protection measures, especially in developing countries, and adaption plans to strengthen preparedness for climatic risks.

Due to the World Economic Forum's Global Risks Report 2022 [3], the results of the recent Global Risks Perception Survey reveal broad concern towards environmental risks. They are among the perceived short-term risks as well as—in the long-term-perspective—the perceived predominant threats and those supposed to be the most severe and most damaging to humans and ecosystems, with "climate action failure" or "extreme weather" as exemplary named forms.

Adaption to and protection from the risks of climate change have economic relevance, revealed in aspects of a wide variety of fields and sectors. Examples of economic fall-outs related to the sphere of adaption to climate risks are shown in the table below (Table 6.1). They point out basic economic rationales in the regarded sphere in an exemplary manner.

TABLE 6.1
Examples of Economic Impacts in the Adaption/Protection Sphere

Possible Economic Impacts	Description
A trade-off between the fast, disordered transition to green economies and decelerated, respectively insufficient measures can be observed.	• With the intensification of physical risks of climate change, the corresponding consequences for societies and economic systems gain urgency. • Challenges of a divisive, disordered, and inequality-creating accelerated transition to carbon net-zero emissions within states and among states worldwide on the one hand, and the risk of slow and surface measures to achieve substantial emissions with devasting outcomes for prosperity and economic development, on the other hand, illustrate the extent of the threat to global and national economies [3].
Strengthening of the domestic preparedness towards climate change risks competes with other future-focused, topical domains, respectively, with the according public and private investments basis.	• Government and business financial aid respectively investments to boost, adapt and protect the domestic societies and economies from climate change fall-outs compete with measures, exemplary, to address pandemics, digitalization, and other future-relevant issues, alongside regular domestic government and private investment programs. • The long-term perspective of climate change adaption aggravates the political and actual feasibility of measures, especially competing with other short- and mid-term domains of critical dimension and high public significance.
Supply-side as well as demand-side economic shocks related to climate change fall-outs endanger economic growth, productivity, work force availability, and private consumption.	• Physical climate change threats like severe weather events and climate disasters can cause economic shocks, which can be embossed by both supply and demand factors [4, 5]. • Agricultural products and raw materials shortfalls as well as damage to production plants can heavily disrupt global and national supply chains and affect the production capabilities of economies. Property damage, business disruptions, and productivity frictions represent serious threats to businesses and the broad economy [6, 7]. Losses of work force can be a further adverse implication in this context [8, 9]. • Ruined homes and livelihoods, respectively the associated monetary losses impair demand of affected households; poor income prospects as well as uncertainty, may lead to less private investments and consumption [4, 5]. A worsening aspect in strongly affected regions consists of potential risk surcharges on loans and insurances or less coverage.

(Continued)

TABLE 6.1 (CONTINUED)
Examples of Economic Impacts in the Adaption/Protection Sphere

Possible Economic Impacts	Description
High recovery and adaption costs and substantial global GDP, as well as human welfare losses with varying degrees of regional impacts, are to be expected in case of minor delivery of climate goals or absent climate action ("business-as-usual").	• From the scientific perspective of the UNFCCC Sixth Assessment Report, limiting global warming requires curbing CO_2 emissions, reaching at least net-zero emissions [1]. Hence, the severity of future impacts of climate change depends on global greenhouse gas emissions, and even ambitious achievements in mitigating the progress of climate change cannot prevent, but at least limit it [10, 11]. • Global GDP impacts of climate change appear to be permanent, long-term and worse with climate action inactivity [12, 10]. Consistently, economic impacts within the adaption and protection sphere are strongly dependent on current and future commitment to climate goals delivering in particular of limiting global warming. • The projected global GDP losses and the particular vulnerability within certain world regions, therefore, hinge on the scope of collective world fulfillment to achieve carbon net-zero emissions. The COP26 Glasgow Climate Pact asks the global community to move forward with NDCs containing ambitious emissions reduction goals committed with the major target to reach net-zero emissions by 2050 [13, 14]. • Examples of rising costs, dropping outputs and economical losses in adversely affected regions are falling productivity in the agricultural sector; infrastructure destruction; decreasing tourism activity, health hazards, and global warming with adverse effects on labor productivity [11, 5]. • Besides lower productivity and reduced work capacity, another channel of amplifying output losses is a reduced investment activity, exemplary due to pessimistic future expectations for the running business or due to damages to capital stock like infrastructure or equipment [5]. Consumer spending is affected adversely in a similar way; for example, as far as worsening future expectations or damages to home infrastructure are concerned [5]. • On trend, adverse economic impacts arise through volatility, pressure on production, and output as well as price hikes and lower future economic growth potential [5].
The disproportionate vulnerability to and the distribution of the consequences of climate change are likely to lead to a redistribution of wealth and economic development worldwide.	• Adverse impacts of climate change, caused damages and losses are observed to affect highly vulnerable people and systems throughout sectors and regions disproportionally, e.g., tropical or low-lying regions [2, 12]. Examples are climate change-related fall-out like droughts or rising sea levels [3]. • Movements of benefits and losses related to climate change among countries and regions causing a turn in comparative (dis-)advantages are supposed to be one of the consequences, exemplary beneficial for the Polar Regions and disadvantageous for the (sub-)tropical regions [11]. This trend does not necessarily depend on the respective region's role as the main harmful emission producer—it rather depends on its particular exposure and susceptibility to the adverse physical impacts of global warming [11].

(Continued)

TABLE 6.1 (CONTINUED)
Examples of Economic Impacts in the Adaption/Protection Sphere

Possible Economic Impacts	Description
Supply chain disruptions might economically impact trade-dependent sectors; countries and regions might be economically triggered by severe weather events, natural disasters, and other appearances related to climatic change.	• Import, export, world markets, global supply chains, product streams, and trade resilience, in general, will probably be affected by changes in climate-sensitive sectors and the corresponding contributing factors, attributes and indicators [8, 11]. Examples can be seen in commodities or agricultural goods, but also manufactured products. • The economic impact potentially rises with the level of dependence of country, region, corporation, organization, etc. on trade on the one hand. On the other hand, the severity of climatic change ramifications on the vulnerably affected systems inclusively the national government's and the domestic private sector's possibilities, as well as the cost-intensity of adaption and protection measures to climate change, play a part [11]. • In this context, there is a significant relevance of the achieved success to expect from the COP26 commitments, especially in the "Adaption" and the "Finance" dimension, in addition also in the "Collaboration" segment.

RESPONSE AND MITIGATION SPHERE OF ECONOMIC IMPACTS

This sphere is characterized by worldwide plans, measures, and action undertaken to deliver climate goals to reduce carbon dioxide emissions and achieve net-zero emissions at least by 2050 [13–15]. The response and mitigation sphere is linked to the major commitment to keep the increase in global average temperature well below 2°C and the pledge to undertake further efforts to limit global warming to 1.5°C [13–15]. The path towards climate goal achievements has been updated at the COP26 Conference in Glasgow, by setting more ambitious 2030 interim goals and an elaborated global system of accordingly adjusted NDCs.

The "Mitigation" segment of the COP26 arrangements, for instance, includes decarbonization and defossilization objectives in the shape of phasing-down coal power or phasing out of fossil fuel subsidies [13, 14].

Not only national governments are addressed by the Glasgow Pact, but, due to its importance towards delivering on global net-zero emissions—in the "Finance" as well as the "Collaboration" segment of the Climate Pact—also the private sector, e.g., with financial institutions or businesses.

A recent McKinsey report [16] classifies sectors primarily responsible of carbon dioxide and methane as climate-warming emissions into seven systems: power, industry, mobility, buildings, agriculture, forestry and waste. Presumably, most economic changes in the regarded sphere are to be expected directly or at least indirectly related to branches of these systems. Thus, they bear particular importance in the mitigation and response sphere underway in a climate-resilient transformation. Table 6.2 outlines possible economic impacts with reference to mitigation and response measures on climate change, especially "green" transitions to limit and to reduce greenhouse gas emissions.

TABLE 6.2
Examples of Economic Impacts in the Mitigation/Response Sphere

Possible Economic Impacts	Description
The transformation process to carbon net-zero emissions is marked by an enormous complexity with polyvalent and pervasive economic impacts on global and domestic economies.	• Due to the UN experts' assessment, an immediate, rapid, extensive, strong, decisive, and sustained approach is necessary to mitigate global warming and climate change [1, 2, 12]—an assessment that formed the basis for the COP26 Glasgow Pact. Enormous investment activity is required to enable a broad transition, from both the public and the private sector, which means immense expenditures [6].
	• The shift to net-zero emissions implicates an epochal global transformation and structural change of global as well as national economies. Besides substantial risks like, for instance, job losses in emission-intensive industries and the fossil fuel-producing sector, stranded assets, climbing energy prices or generally shifting prices, and energy supply volatility [16, 17], significant opportunities and promising perspectives can be expected of a green transformation and climate-resistant development. Such opportunities could be newly created jobs, efficient defossilization processes, low-emission products, or new, sustainable technologies e.g. in power generation, transport and agriculture [18, 19, 9].
	• The extraordinary transformation requires extensive capital spending e.g. on assets, processes, and supply chain adaptions of greenhouse gas-emitting systems [16, 17], an outlook that underlines the important role of the private sector in the mitigation sphere. This aspect is aligned with the systems-thinking perspective [4, Part I], where different nuances can be factored into and be a substantial part of a process, thereby enmeshing technical, social, political, economic, cultural, and managerial considerations.
	• A certain disorder in the execution of climate transition despite common goals is to be expected: cost increasing and losses of working places in carbon-intensive economic sectors, missing out on the technological development or rising inequality could become unexpected concomitants, reinforcing frictions and undermining acceptance [3].
	• The economic risks of complexity have the chance to become manageable through intensive and broad cooperation. The Working Group II contribution to the IPCC Sixth Assessment report [2] points out that climate-resilient development is effectively promoted by inclusive governance, decisions, and actions, involving national governments, civil societies, and the private sector as well as international cooperation and partnerships with traditionally marginalized groups.

(Continued)

TABLE 6.2 (CONTINUED)
Examples of Economic Impacts in the Mitigation/Response Sphere

Possible Economic Impacts	Description
Carbon trading and pricing are likely to become a main part of instruments to manage climate change mitigation—they affect the government sphere, businesses as well as consumers economically.	• Carbon dioxide pricing, e.g. taxes, transparently fixes prices typically for a certain level of CO_2 [18]. Hence, it raises prices and reduces the profitability of carbon-footprinted products and services, respectively tending to set incentives for carbon dioxide-emitting systems as well as private consumers to shift to climate-friendly products and services. Examples are to be found in heating systems, high-emission foods, or vehicles.
	• COP26 achievements denounce coal as the most polluting of the existing fossil fuels; moving away from coal power, though, could reinforce certain global economic imbalances.
	• Exemplarily, the acknowledgment that industrialized nations are much more responsible for the climate problem could force developed economies to suspend their cheap fossil fuels on the one hand and encourage a persistent usage of inefficient fossil-fuel systems of coal-dependent, growing developing nations on the other hand [7, 19]. In a globalized world, this trend could lead to competitive disadvantages and distortions, especially if occurring in combination with financial burdens companies already need to deal with, e.g. from carbon emissions trading [19].
	• Carbon emission trading corresponds to a carbon cap system, which allows the transfer of carbon dioxide emission "credits" to other market participants, charging over polluters and rewarding those who can reduce emissions below the permitted limit [18]. Economic opportunities potentially arise from the relatively low costs to societies as a whole as well as the political feasibility [18], moreover, the incentive-creating approach to stimulate the market's climate transformation efforts. Carbon trading charges certain market participants and disadvantages, for instance, financially weak, less climate-innovative, emission-intensive, respectively, fossil fuel-dependent businesses, industries, and sectors.
Economic shocks and dislocations can also become accompanying effects in the mitigation sphere.	• Economic shocks cannot only be caused by physical impacts such as severe weather events or natural disasters. They can also be triggered by a tight low-carbon climate policy, e.g. causing disruptions and dislocations in certain carbon-intensive lines of business [4].
	• Fossil fuel subsidies elimination or carbon taxes are examples of installed response policies to mitigate climate change with a potentially high adverse impact on the competitiveness, economic contribution, workforce, etc. of intensively carbon emission polluting sectors [11].

(Continued)

TABLE 6.2 (CONTINUED)
Examples of Economic Impacts in the Mitigation/Response Sphere

Possible Economic Impacts	Description
Cross-border trade aspects linked to climate change mitigation could also include economic fall-outs.	• Implemented climate standards or requirements on products and for production, exemplary comparable to safety standards for the engagement and the protection of the labor force in manufacturing processes, have the potential to lead to cross-border trade impacts [7, 11]. Examples include introduced climate standards on certain products or border CO_2 taxes. • Impacts of response policies to climate change on trade and economic development in this context are complex to estimate, in general, and depend on several connected factors such as transportation costs or competitive advantages of certain sectors [11]. • Nevertheless, trade-related mitigation measures could also lead to incentives, and a "drive" in moving forward to meet emission limiting requirements. On the other hand, they carry economic risks for regions, companies and the workforce facing affection by rising costs or shrinking revenues.
Public investments compete with private investments in a low-carbon transition. In this context, COP26 achievements are expected to have a boost effect on the efforts by the private business sector to net-zero emissions.	• Adapted NDCs, a major part of COP26 results, tend to strengthen national ambitions to align strategies, initiatives, endeavors, and policies with the net-zero goal. • In general, not only political negotiations but also impulses and pushes on and from the private sector towards a climate-resistant transition were an important part of the COP26 agreements (collaboration segment, a.o.). • Decarbonization could probably be based on different investments in an imbalanced way—a primary government investment could crowd out private sector investments [4]. • However, reassessed business and government strategies taking into account carbon balances, increasing pressure on emitting sectors, and on policies to realign on more ambitious climate action inclusively climate-related investments could gain significantly in importance [19]. This is why political endeavors, organizations, or businesses without an affiliated low-carbon plan could be affected adversely, related to credit ratings, risk as well as insurance assessments or attractiveness to talents etc. [19]. Examples are waiver of air traffic, rail or public transport promotion measures; usage of renewable energy sources, and zero-emission duty vehicle fleets.

SELECTED BUSINESS ASPECTS, INSTRUMENTS AND TOOLS

THE ECONOMIC ANALYSIS BASIS FOR SYSTEMS INTEGRATION

In the final analysis, all organizations are interested in the bottom line (i.e., the economic basis) of their operations. In this regard, we provide additional perspectives in this section. "Evaluation" is an essential part of the DEJI Systems Model, beyond the "Design" phase. This contains an engineering-technical focus, through which the DEJI Systems Model provides an economic evaluation of the overall system, with a particular emphasis on the systems-integration requirement.

The evaluation stage mainly includes the economic feasibility of the regarded system within a System-of-Systems approach. This stage might be supplemented, accompanied, respectively expanded by a cost–benefit analysis in the meaning of a **utility dimension**. The so-called **cost–benefit analysis** can be used as a supplement to conventional monetary profitability calculations, especially when two or more options have almost the same characteristics and evaluation variables.

COST–BENEFIT ANALYSIS (BENEFIT = UTILITY): MICROECONOMIC PERSPECTIVE

As a case example, suppose a corporation in an economic sector of the critical infrastructure dimension (shown in Figure 6.1) is planning to conduct an investment into a production line connected to an autonomous energy-generating and energy-providing system. The company has carried out the value of investment costs for three different system options, including the remaining values at the end of the period of consideration. We assume that the "classic" investment accounting based on the capital-value method [6] yields preliminary results along with the fact that no clear decision can easily be made. The net present values and internal rates of return are, in this case example, only marginally different. After carrying out the first part of the economic analysis, the investment cost perspective, the available production line systems are analyzed on the value of their utility contribution as well as benefits from avoided costs. The techniques of utility modeling can be used to assess the relative utilities of

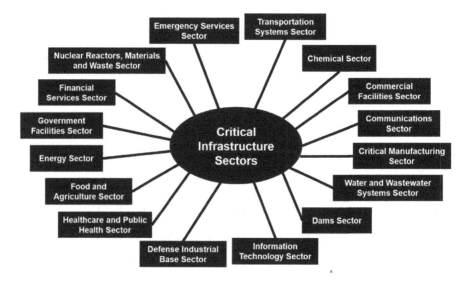

FIGURE 6.1   The 16 Critical Infrastructure Sectors in the United States (Based on [21]).

TABLE 6.3
Considerable Benefits to Be Measured in a Cost–Benefit Analysis for the Case Example

Reduction of lead/turnaround/processing times
Reduction of scrap and error rate
Strengthening of strategic cooperations and partnerships with associated organizations
Improvement of delivery readiness
Improvement of work processes and working conditions
Amplification of employee skills, knowledge, and capabilities
Improvement of service or product quality and reduction of penalty costs related to quality
Enhancement of the customer satisfaction and reduction of customer complaints
Avoidance of fossil fuels and enhancement of the use of renewable energy sources
Improvement of employee motivation and raise of attractiveness to talents
Improvements in communication and information flow
Improvement of the company image and reputation
Reduction of overtime and avoidance of extra work
Raise of autonomy and the company's independence from suppliers
Reduction of energy consumption, saving of energy costs, increase in energy efficiency
Improvement of the adherence to delivery dates
Reduction of internal transport routes and times

sub-elements in the system (Badiru and Foote, 1992 [20]). Table 6.3 illustrates examples of benefits considered in the economic assessment of our hypothetical case.

An example of a beneficial outcome within this microeconomic perspective is the potentially positive reputation effect of an emissions-neutral and climate-conscious investment option, connected with an autonomous energy system that is based on renewable energy sources. With the aid of the cost–benefit analysis, thus, the climate-related perspective can be easily enmeshed and evaluated within the microeconomic sphere of an organization. The expected "green" impacts can even be pertinently isolated and measured [22]. In the hypothetical study, reinforcement of energy efficiency, raising the awareness of the potential workforce, and increasing the independence of workers are further exemplary beneficial factors regarding climate change. As outlined in Table 6.2, climate change mitigation takes hold in the microsphere of a company, a.o. Hence, reassessed business strategies taking into account carbon emission balances, respectively, climate-related action and investments have the potential to increase in significance in this economic dimension.

In the case example, the utility of the new energy system, quantitatively measured in benefits, can be related to the investment costs of the respective production line with its connected energy system and incorporated into an extended profitability analysis—extended by the sum of total annual benefits over the period of consideration of the investment case. A major prerequisite is the requirement to convert identified benefits into a monetary value, based on comparable units [22]. The result of the cost–benefit analysis should be, in the above-mentioned example, the deciding

TABLE 6.4

Overview: Exemplary Cost–Benefit Calculation using the Hypothetical Case

	Option A	Option B	Option C
$NPV = \sum_{n=1}^{T} \dfrac{CF_n}{(1+r)^n}$	25,000,000	24,500,000	25,800,000
$VB = \sum_{n=1}^{T} \dfrac{B_n}{(1+r)^n}$	12,000,000	4,500,000	6,000,000

factor due to the similar net present values and the significant differences of benefit factors (see Table 6.4). Besides, the approach bears "bottom-up" importance as a sort of microeconomic perspective on climate change mitigation, as revealed above. In the case example, we suppose Option A has the highest overall efficiency due to the leading cost–benefit relation in comparison to the competing options, hypothetically foremost regarding climate-related benefits.

Such a cost-benefit analysis can, in general, be performed for various areas of application. There are, however, several challenges to deal with. For example, due to different expectations regarding the particular benefits of measures. One further characteristic of the cost-benefit analysis, according to the fact that values of benefits have to be carried out, is the necessity of comparable utility units among different options and therefore the most possible objectification and also most precise and factual possible monetarization of the respective benefit aspects.

Where
  NPV describes the net present value of the respective investment option,
  CF is the annual net cash flow forecasted for each year of the period of consideration (defined by cash inflows – cash outflows in the respective year),
  n is the year of the annual cash flow or of the total annual benefits,
  T is the period of consideration of the investment project (in years),
  r is the discount rate,
  VB describes the discounted value of aggregated benefits of the respective option,
  B is the predicted monetary value of the total annual benefits for year *n*.

In almost every investment project, there are non-monetary factors that are not considered or only marginally considered in classic profitability analyses. Among other things, this approach enables us to assess an investment, system, or project more comprehensively and completely. This is one of the beneficial efficacies of the DEJI Systems Model. It may turn out after the evaluation that a project that has the greatest advantage from a monetary point of view is not the most beneficial. A weighting, for example, which considers 80% of the results of the monetary profitability analysis and 20% of the results of the cost-benefit analysis, helps to objectify the results. This is an example of the classic Pareto Distribution Analysis of investment projects.

COST–BENEFIT ANALYSIS: MACROECONOMIC PERSPECTIVE

Due to its comprehensive character, the cost-benefit analysis involves an appropriate procedure for the assessment of macroeconomic efficiency, e.g. of government policies or law initiatives. Concerning the COP26 pledges, adapted NDCs are the means to express national contributions and actions to reduce emissions, deliver on goals, and mitigate climate change. On this macroeconomic level, the cost-benefit analysis also allows incorporating the beneficial outcome of climate-related measures. Benefits are part of an extensive approach to efficiency assessment. They can be pertinently related to the monetary perspective on legislation and public investments and therefore supplement the wide-ranging evaluation of the impacts of measures.

In contrast to the microeconomic approach described above, in this case, a macroeconomic view on measures is adopted, i.e., all positive and negative effects of the measure are taken into account, irrespective of the reason for their occurrence.

Depending on the degree of detectability and the possibility of monetarization, the effects which can be considered in the analysis involve:

- *Direct costs*, respectively, *revenues* that can be directly determined based on available market prices. Exemplarily, we can assume, for instance, investment costs of a railroad and railway station construction and permanent revenues from the use by railroad companies.
- *Indirect costs*, respectively, *benefits* that can be monetarized through comparative estimates. Suppose e.g., the increased noise pollution caused by the station and the railroad traffic; the strengthening effect on the city's attractiveness to new residents and purchasing power, or the climate-related impact of emission avoidance.
- *Non-monetizable costs*, respectively, *benefits* that can be evaluated through an advantage and disadvantage illustration or a pure utility value analysis. Examples, in this case, are changes to the landscape by the railroad station, or a modern construction design.

An important prerequisite for the use of cost-benefit analysis, especially to ensure comparability, is a set of specifications including the respective objectives, measure effects to be captured, the evaluation scale, etc. Monetizable costs and benefits occurring at different times have to be taken into account with their net present values (see methodology of the microeconomic perspective).

A challenge with particular regard to the involvement of climate change mitigation in the conduct of macroeconomic cost-benefit analyses is to quantify the impacts of climate-related policies. As these impacts have a temporal aspect, and, furthermore, often no valuable market prices exist to evaluate them, it is necessary to connect them with the physical impact as well as the mitigation impact perspective of the regarded world region and nation. Thus, the described spheres (adaption and protection, mitigation and response) with the according expected economic and environmental impacts inclusively trade-offs and side effects, are relevant for the macroeconomic cost-benefit assessment of measures. The methodology is similar to the microeconomic perspective—measurable, comparable units form the basis for a calculation of

TABLE 6.5

Examples of Measurable Impacts in a Cost-Benefit Analysis
(Macroeconomic View)

Quantified environmental damages and adverse economic fall-outs saved by certain protection policies
and investments

Influence on crop yield security by measures which are installed to curb physical threats resulting from
climate change

Respective avoidance of carbon dioxide emissions through a cut of fossil fuel subsidies in relation to
the loss of workforce in the fossil fuel sector

New labor opportunities related to a carbon net-zero transition of the energy generation sector

Technological progress emerging through shifting to emission-reducing means of transport, waste
processing and recycling, or manufacturing technologies

Consideration of threat-minimizing, emissions-reducing and renewable energy-producing effects in
infrastructure construction and restructuring projects

Public profits from reinvestments of carbon revenues in return for impacts from carbon pricing and
energy costs on demand and growth [5]

Fossil fuel energy savings from power line investments in renewable energies

monetary climate-related factors. In addition, non-monetizable factors are supple-
mentarily assessable on a utility value basis.

The dimension of a macroeconomic perspective on cost-benefit analyses is probably more complex than the microeconomic one. To carry out analyses of this dimension requires fundamental data sources and a set of specifications, as mentioned
above. On the other hand, this instrument is, for instance, appropriately capable of
contributing to an assessment of measures in the context of national contributions to
climate change adaption and mitigation. Its methodology enables taking climate
change aspects into consideration in national legislation, initiatives and policies, as
well as sharpening political arguments. Examples of measurable and evaluable
impacts in this context are outlined in the table below (see Table 6.5).

All in all, the macroeconomic cost-benefit approach allows considering quantified avoided costs, prevented environmental damages, and other benefits attributable
to policies, and also to relate economic impacts to climate change mitigation strategies. Without a pervasive quantification of benefits of domestic policy endeavors, it
will become more difficult to perform a successful climate change adaption and
transition to achieve climate goals policy measures, and to convince stakeholders
and the population.

LIFE CYCLE COST MANAGEMENT

Life cycle cost management is a further methodology that allows the incorporation of
economic impacts of climate change, the adaption as well as the mitigation dimension. In general, the **life cycle costing** approach is a means to estimate the entire costs
of products, services or work processes. It includes and takes into account the whole
life cycle of products etc., throughout the whole life dimension "from the cradle to
the grave."

TABLE 6.6
Examples of Life Cycle Cost Considerations in the Context of Economic Impacts from Climate Change (Domain of Business Decisions, Costs of Products/Services/Processes)

Impacts of supply shocks as raw materials, agricultural goods shortfalls, and energy prices risings

Investments in sustainability, defossilization, climate resilience, technological progress, emission mitigating transition as additional expenditures/further apportionment costs on products/services

Climate risk (physical as well as transition risks) and venture cost reconsideration

Additional disposal costs for environmentally harmful waste and carbon emission-intensive product removals

Finance costs/interest rates reconsideration: e.g. risk surcharges on insurance fees and credit costs

Consideration of underinsurance and lack of coverage in case of significant physical climate risks

Losses, damage, and deprecation in capital stock/assets according to climate change threats

Carbon pricing and carbon trading

Affecting changes in public subsidies (e.g. for fossil fuels) and government priority adaption

Prolonged stock holding due to supply chain disruptions

Cross-border trade impacts related to climate change, such as product climate standards

Affection of labor force productivity and outages caused by severe weather events like heat waves

In the domain of business decisions, it is conducive to evaluate and to consider the total product prices comprehensively, e.g. in the case of a new asset investment for the manufacturing process of a product. In this domain, many economic impacts of climate change (see Table 6.1, Table 6.2), which directly affect the cost consideration of the respective company for the particular investment project, can be enmeshed. Examples are shown in Table 6.6 (aligned with [17]).

ROI/EVA—SUPPLEMENTARY DIMENSION

Climate change measures are, to a certain extent, comparable to regular business investments: they require financial efforts and provide benefits, mostly long-term, due to the strengthening of preparedness to tackle physical threats of climate change and due to the transition to achieve emission reduction goals. This is why climate-related investments help to avoid future economic losses on the hand, respectively, to develop climate-resistant economic progress, profiting from opportunities of decarbonization. Climate action measures require investments, which ideally "pay off" in a lifecycle approach by providing substantial benefits. The mathematical equation for the **Return on Investment** (ROI), which is useful in this context, is to be described as follows:

$$\mathbf{ROI} = \frac{\text{Net investment gain / loss}}{\text{Cost of investment}} \times 100$$

$$= \frac{\left(\text{Current value of investment} - \text{initial costs of investment}\right)}{\left(\text{Initial costs of investment}\right)} \times 100$$

To measure, track and compare efficiency of "green" investments, a *climate-related ROI* can be derived from the above-mentioned principle to assess climate-related investment benefits:

$$Climate\text{-}related\ ROI = \frac{\left(\text{Net Benefits from climate investments}\right)}{\left(\text{Costs of climate investments}\right)} \times 100$$

Even though certain aspects such as the time dimension of the investment or different types of costs are excluded from this approach, it nevertheless provides an additional feasible option to assess climate investment initiatives under the application of an economic view. As in the other introduced instruments, the supplementary ROI dimension also requires measurable units of gained benefits, with the help of which benefits from the contribution containing climate change can be translated into a value.

A possible alternative, with generally a similar implication, methodological principle, and meaning to evaluate and compare efficiency of climate action spending and measures, is the **Economic Value Added** (EVA) approach. The equation for the EVA can be described as follows:

$$EVA = \text{Net operating profit (After Taxes)} - (\text{weighted average cost of capital} \times \text{capital invested})$$

A supplementary view on the abovementioned instrument could involve net benefits gained from climate-related investments and set them into relation to the costs of the accordingly invested capital:

$$Climate\text{-}related\ EVA = \text{Net} \times \text{benefits from climate investments} - (\text{cost of capital invested} \times \text{capital invested})$$

In simple terms, the climate-related EVA describes whether the value of generated benefits exceeds the cost of invested capital. This approach delivers a contribution to the comparability of climate-relevant expenses and the gains of different organizations as well as of different periods. Thus, it allows, in a particular way, a measurement of the efficiency of investments meant to protect from and to respond to climate change. Furthermore, the EVA instrument draws upon its economic origin to measure the capability to generate additional value beyond the costs of invested capital, but it is certainly not limited to a financial view on businesses. The principle applies to a variety of cases, domains, and organizations facing and operating climate action investment cases.

TARGET AGREEMENTS

From the economic perspective, a further potentially appropriate tool to deliver, manage and break down climate goals is worth considering. COP26 agreements, pledges, and commitments and the derived NDCs could be operationalized by pouring them

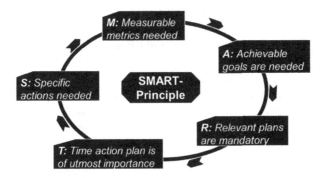

FIGURE 6.2   SMART principle of operations management.

into the molds of target agreements and sub-targets in politics, business, financial institutions, private organizations, and administration according to the SMART principle of operations management.

We remember, as previously mentioned, a systems approach facilitates endeavors following the SMART principle of operations management, which is shown in Figure 6.2.

The application of the SMART principle of operations for the purpose of target operationalization and also as tracking management within different target spheres, top-down measured, is a possible vehicle to strengthen domestic efforts of national climate ambitions. Without an appropriate operationalization and tracking system, the most ambitious global climate goal is worthless—the "drive" created by national impulses from governments, the private sector, and societies can only accelerate, if COP26 pledges are imbued with life by global stakeholders. The Systems Modeling view is of the highest relevance in this context: divergent national platforms of political systems, cultural systems, legal systems, financial systems, educational systems, media systems, trade systems, etc., require these dissimilarities and disconnections to be appropriately addressed. Hence, the integration stage of Systems Modeling considering cultural and operational differences and accompanied by accordingly operationalized climate goals, allows an effective and efficient moving forward as far as climate change action is concerned.

Target agreements based on the SMART principle of operations can be installed in almost every hierarchical sphere. Examples can be found in politics (e.g. Chief of Government, Secretaries, Parliamentary State Secretaries), business (e.g. CEO, COO, Heads of Departments, subsidiaries), and administration (e.g. State Secretaries, Heads of Directorates, Heads of subordinate agencies). A highly relevant part of target agreements is an effective tracking and controlling system, cutting paths into the operations of different organizations. Moreover, incentives and consequences in case of achievement, under-fulfillment, and failure to reach goals complete a pertinent establishment of target agreement systems.

Table 6.7 illustrates examples of the contents of operationalized national target agreements.

TABLE 6.7

Examples of Contents of Operationalized Climate-Related Target Agreements

Targets on zero-emission duty vehicle fleets

Zero-carbon-oriented procurement goals

Defossilization cooperation goals (e.g. joint ventures, supplier alliances)

Guidelines and standards of usage of renewable energy sources in manufacturing processes

Emission reduction goals for duty transportation and traveling (e.g. waiver of air traffic and promotion of rail and of public transport)

Training and education goals to enhance acceptance and consciousness

Goals on system thinking, comprehensive cooperation with civil society, educational bodies, and scientific institutions in climate-related aspects, underinclusion of marginalized groups

Prescriptive consideration of climate aspects in negotiations, contracting, and trade

Goals on emission-reducing public construction measures and government investments

Installed representatives/persons in charge of climate change issues within organizations

Prescriptive implementation into decision-making processes and consideration in guidelines

SYSTEMS EFFICACY

Without a systems view, we cannot be sure we are pursuing the right outputs that can be integrated into the prevailing operating environment. Thus, the DEJI Systems Model allows for a multidimensional analysis of climate matter considering many of the typical "ilities" attributes related to a System-of-Systems view. It focuses on specific goals of a system considering the specifications, prevailing constraints, possible behaviors, and structure of the system, and involves a consideration of the activities required to ensure that the system's performance matches specified goals. Concretely, this means the measures and activities to fulfill global climate objectives, having been reaffirmed at COP26, passed through a system engineering perspective and its particular integrative process methodology on a stage-by-stage approach. Thereby, the methodology contains an interdisciplinary assessment of the legal, economic, and technical nuances impinging on the viability of the implementation of the pledges and premises that emanated from COP26 in Glasgow, Scotland in 2021, particularly related to the thematic dimension of critical infrastructure.

LEGAL ASPECTS

The implementation of policies regarding decarbonization is conducted by the implementation of specific regulations. This is especially true for the promotion of new technologies, e.g. carbon capture and storage/removal and geoengineering. In addition, different aspects of administrative, civil, constitutional and (international) trade law need to be considered when implementing further policies like the adjustment of public procurement procedures, the adoption of a mandate to disclose climate-related risks, and the imposition of tariffs on imports. Some of the identified open questions have not yet been subject to any legal proceedings. Hence, no respective precedents exist. This might lead to new litigation cases but also creates an opportunity to conduct further research on specific issues from a legal point of view.

ECONOMIC ASPECTS

All in all, there are economic risks and threats related to a transformation to carbon net-zero emission, especially to jobs, industries and sectors with a particular emission intensity and reference to fossil fuel. The global economic structure will certainly change significantly. But on the other hand, costs and GDP losses in a business-as-usual or delayed transition scenario in which the global community fails to undertake necessary efforts to respond to climate change and to deliver climate goals, have the very likely potential to cause extensive adverse fall-outs to economies and irreversible damages to human life. The COP26 Climate Pact, with its various segments, involves a promising toolbox to deal with both the adaption as well as the mitigation sphere, so that the states and their decision-makers in governments, private organizations, financial institutions, industries, and companies have been provided with an effective perspective on an ambitious pathway to curb the climate change-related threats to their societies and economies. The effectiveness of decisive climate change policies is essentially supported by strong international cooperation as well as a comprehensive, collaborative approach between societies, governments, and the private sector perpetually including traditionally marginalized population groups. In addition, there are not least economic opportunities with regard to climate transitions, awaiting those who will move forward to respond to climate change in the framework of a "green" transformation.

In this context, the methodology of various economic techniques and tools is strongly applicable to the systems view on climate change agendas, initiatives, and policies presented in this chapter. The supplementary consideration of these business aspects allows the measurement of the efficiency of climate action measures and therefore involves the capability to accompany the development of protection against climatic threats and mitigation, respectively, "green" transition towards carbon net-zero.

In correspondence with the systems-thinking perspective of Systems Engineering different nuances, consisting of technical, social, political, economic, cultural, and managerial considerations, among others, are factored into and form a substantial part of a System-of-Systems view on a complex process, such as climate change adaption and mitigation based on COP26 is.

CONCLUSIONS

Outputs of System-of-Systems follow an integrative process that must be evaluated on a stage-by-stage approach. This requires research, application, and implementation strategies that consider several pertinent factors [23, 24, 25–41]. This collaborative chapter suggests the DEJI Systems Model, which has been used successfully for product development applications, as a viable methodology for system design, system evaluation, system justification, and system integration. The integrative approach of the DEJI Systems Model can facilitate the intricate and complex interrelation between climate change, COP26 Glasgow Climate Pact agreements as well as critical infrastructure by application of a systems view and consideration of existing analytical tools and techniques. The systematic involvement of additional technical-engineering, legal and economic perspectives respectively aspects as parts of the systems approach represent the main focus and output. The chapter reveals that the climate change resilience endeavor is coherently combinable with the structured framework of Systems Modeling. It furthermore stresses the alignment to the scope of practical systems implementations in business, industry, government, financial institutions, and private organizations, among others.

It is anticipated that this Joint Journal Paper sparks the interest of researchers in the DEJI Systems Model and emphasizes its polyvalent problem-solving applicability on multifaceted problem domains like climate change and critical infrastructure, so that collaborative research can help advance the mathematical and graphical representation of the model elements, with a special focus on the metrics of complex systems integration. It is moreover hoped to provide a pertinent platform to contribute to the climate change debate in a global resolve context.

Such an application will expand the applicability of the DEJI Systems Model to platforms of national debates related to climate change mitigation initiatives. Hence, it can be even applied to evaluate and justify a designed politically-driven national proposal, with a specific template for integration into the expectations and peculiarities of the local or regional environment.

REFERENCES

1. IPCC (2021), Sixth Assessment Report, Working Group I: The Physical Science Basis; Headline Statements from the Summary for Policymakers, August 9, 2021. https://www.ipcc.ch/report/ar6/wg1/ (accessed 23 July 2022).
2. IPCC (2022), Working Group II Contribution to the Sixth Assessment Report of the IPCC: Climate Change 2022, Impacts, Adaption and Vulnerability; Headline Statements from the Summary for Policymakers, February 28, 2022. https://www.ipcc.ch/report/ar6/wg2/ (accessed 2 October 2022).
3. World Economic Forum (2022), *The Global Risks Report 2022*, 17th Edition, ISBN: 978-2-940631-09-4. https://www.weforum.org/reports/global-risks-report-2022
4. Batten, S., R. Sowerbutts, and M. Tanaka (2020), "Climate Change: Macroeconomic Impact and Implications for Monetary Policy", in: *Ecological, Societal, and Technological Risks and the Financial Sector*, Thomas Walker, Editor, Palgrave Macmillan, Cham, July.
5. Andersson, M., C. Baccianti, and J. Morgan (2020), "Climate Change and the Macro Economy", published with Occasional Paper Series, European Central Bank, No. 243, June 2020.
6. Vanek, Francis M., Louis D. Albright, and Largus T. Angenent (2016), *Energy System Engineering: Evaluation and Implementation*, Third Edition, McGraw-Hill Education, New York, NY.
7. United Nations Conference on Trade and Development (UNCTAD) (2021), *Trade and Environment Review 2021: Trade-Climate Readiness for Developing Countries*, New York, ISBN: 978-92-1-113009-6.
8. Badiru, Adedeji B. (2022), *Global Supply Chan: Using Systems Engineering Strategies to Respond to Disruptions*, Taylor & Francis Group/CRC Press, Boca Raton, FL.
9. Badiru, A. B., and B. L. Foote (1992), "Utility Based Justification of Advanced Manufacturing Technology," in: *Manufacturing Research and Technology* (Vol. 14, pp. 189–207), H. Parsaei, W. Sullivan, and T. Hanley, Editors. Elsevier, New York, NY.

10. Nair, S. (2021), "Climate Inaction Costlier than Net Zero Transition: Reuters Poll", Reuters, October 25, 2021 https://www.reuters.com/business/cop/climate-inaction-costlier-than-net-zero-transition-economists-2021-10-25

11. United Nations Conference on Trade and Development (UNCTAD) (2021), *Climate Change, Green Recovery and Trade*, New York, eISBN: 978-92-1-005630-4.

12. John, M. (2021), "COP26, Explainer: Climate Change: What Are the Economic Stakes?", Reuters, November 5, 2021. https://www.reuters.com/business/cop/climate-change-what-are-economic-stakes-2021-10-25

13. United Nations/United Kingdom Government (2021), "UN Climate Change Conference UK 2021", COP26 – The Glasgow Climate Pact, https://ukcop26.org (accessed April 19, 2022).

14. United Nations (2022), "Climate Action – COP26: Together for Our Planet", https://www.un.org/en/climatechange/cop2 (accessed April 25, 2022).

15. United Nations Framework Convention on Climate (2021), "Glasgow Climate Change Conference: October–November 2021", (available on: https://unfccc.int/conference/glasgow-climate-change-conference-october-november-2021 (accessed April 19, 2022).

16. ILO – International Labour Organization (2022), "Green Jobs – Frequently Asked Questions on Climate Change and Jobs", https://www.ilo.org/global/topics/green-jobs/WCMS_371589/lang--en/index.htm (accessed May 3, 2022).

17. Guidance by the Office of Management and Budget (OMB) (n.d.), "OMB Circular A-4". www.whitehouse.gov/wp-content/uploads/2023/04/DraftCircularA-4.pdf, (accessed April 5, 2023).

18. Pappis, Costas P. (2011), *Climate Change, Supply Chain Management and Enterprise Adaption: Implications of Global Warming on the Economy*, IGI Global/Information Science Reference, Hershey, PA.

19. Jessop, S., J. Spring, and R. Kerber (2021), "Analysis: COP26 Message to Business - Clean Up to Cash In", Reuters, November 14, 2021. https://www.reuters.com/business/cop/cop26-message-business-clean-up-cash-2021-11-14/

20. Network for Greening the Financial System (NGFS) (2021), "Climate Scenarios for Central Banks and Supervisors", June 2021. https://www.ngfs.net/en/ngfs-climate-scenarios-central-banks-and-supervisors-june-2021 (accessed June 5, 2022).

21. Cybersecurity & Infrastructure Security Agency (CISA) (2022), Infrastructure Security: Critical Infrastructure Sectors, https://www.cisa.gov/critical-infrastructure-sectors (accessed April 14, 2022).

22. Williams, Patricia P. and Jack J. Phillips (2011), *The Green Scorecard: Measuring the Return on Investment in Sustainability Initiatives*, Nicholas Brealey Publishing, Boston, MA, USA/London, UK.

23. Badiru, Adedeji B. (2019), *Systems Engineering Models: Theory, Methods, and Applications*, Taylor & Francis Group/CRC Press, Boca Raton, FL.

24. McKinsey (2022), "The Net-Zero Transition: What It Would Cost, What It Could Bring", https://www.mckinsey.com/business-functions/sustainability/our-insights/the-net-zero-transition-what-it-would-cost-what-it-could-bring (accessed April 26, 2022).

25. Fact Sheet: Executive Order Catalyzing America's Clean Energy Economy Through Federal Sustainability December 8, 2021. https://www.whitehouse.gov/briefing-room/statements-releases/2021/12/08/fact-sheet-president-biden-signs-executive-order-catalyzing-americas-clean-energy-economy-through-federal-sustainability/ (accessed December 8, 2022).

26. A European Green Deal (n.d.), https://ec.europa.eu/info/strategy/priorities-2019-2024/european-green-deal_en (accessed May 12, 2022).

27. National Academies of Sciences, Engineering, and Medicine (2021). *A Research Strategy for Ocean-based Carbon Dioxide Removal and Sequestration.* The National Academies Press, Washington, DC. https://doi.org/10.17226/26278 (accessed May 12, 2022).

28. Douglas, Leah (2022), "U.S. Carbon Pipeline Proposals Trigger Backlash over Potential Land Seizures", Reuters, February 7, 2022. https://www.reuters.com/business/environment/us-carbon-pipeline-proposals-trigger-backlash-over-potential-land-seizures-2022-02-07/ (accessed February 7, 2022).

29. Benson, Thor (2022), "How to Weaponize Our Dying Oceans against Climate Change", The Daily Beast, February 14, 2022. https://www.thedailybeast.com/how-the-ocean-could-be-the-key-to-carbon-capture-technologies-to-fight-climate-change, (accessed February 14, 2022).

30. Joselow, Maxine (2022), "The SEC Will Propose a Historic Climate Disclosure Rule. Here's What to Know", The Washington Post, March 15, 2022. https://www.washingtonpost.com/politics/2022/03/15/sec-will-propose-historic-climate-disclosure-rule-here-what-know/ (accessed March 15, 2022).

31. Walters, Daniel E. and William M. Manson (2022), "SEC Will Consider Climate Disclosure Rules for US Companies on March 21 – It's Already Facing Threats of Lawsuits", The Conversation, March 7, 2022. https://theconversation.com/sec-will-consider-climate-disclosure-rules-for-us-companies-on-march-21-its-already-facing-threats-of-lawsuits-178304 (accessed March 22, 2022).

32. Rana, Ferhan (2022), "SEC's Climate Disclosure Proposal Would Force Firms to Tell Investors the Truth about Emissions", TrivDaily, March 22, 2022. https://trivdaily.com/secs-climate-disclosure-proposal-would-force-firms-to-tell-investors-the-truth-about-emissions#:~:text=Home%20Tech-,SEC's%20Climate%20Disclosure%20Proposal%20Would%20Force%20Firms,Investors%20the%20Truth%20About%20Emissions&text=Companies%20looking%20to%20trade%20on,related%20threats%20to%20potential%20financiers (accessed March 22, 2022).

33. Weise, Elizabeth (2022), "How Will Climate Change Impact American Companies? The SEC Thinks You Have a Right to Know", USA Today, March 20, 2022. https://www.usatoday.com/story/news/2022/03/20/new-sec-rule-companies-disclose-climate-change-risks/7048413001/?gnt-cfr=1 (accessed March 20, 2022).

34. The Week (2022), "What Is Carbon Capture and Storage?", April 5, 2022. https://www.theweek.co.uk/news/environment/956334/what-is-carbon-capture (accessed April 5, 2022).

35. Ridley, Kirstin and Simon Jessop (2022), "Analysis-as Tougher Climate Disclosures Hit, Campaigners Scour for Laggards", Reuters, April 7, 2022. https://www.reuters.com/business/environment/tougher-climate-disclosures-hit-campaigners-scour-laggards-2022-04-07/, (accessed April 7, 2022).

36. IPCC (2022), "Climate Change 2022: Impacts, Adaptation, and Vulnerability. Contribution of Working Group II to the Sixth Assessment Report of the Intergovernmental Panel on Climate Change", H.-O. Pörtner, D. C. Roberts, M. Tignor, E. S. Poloczanska, K. Mintenbeck, A. Alegría, M. Craig, S. Langsdorf, S. Löschke, V. Möller, A. Okem, and B. Rama, editors Cambridge University Press.

37. Kusnetz, Nicholas (2022), "Carbon Capture Takes Center Stage, but Is Its Promise an Illusion?", Inside Climate News, March 9, 2022. https://insideclimatenews.org/news/09032022/carbon-capture-and-storage-fossil-fuels-climate-change/?gclid=EAIaIQobChMI1aKr95jB9wIV8waICR3urgBpEAAYASAAEgK5HPD_BwE, (accessed March 9, 2022).

38. Frazin, Rachel (2022), "Lawmakers Consider Carbon Border Tax, Environmental Reviews at Bipartisan Climate Meeting", THE HILL, May 2, 2022. https://thehill.com/policy/energy-environment/3474664-lawmakers-consider-carbon-border-tax-environmental-reviews-at-bipartisan-climate-meeting/ (accessed May 2, 2022).

39. Frazin, Rachel (2022), "Oversight Republicans Target SEC Climate Disclosure Proposal", THE HILL, May 4, 2022. https://thehill.com/policy/energy-environment/3477023-gop-targets-proposed-rule-requiring-companies-to-disclose-climate-change-contributions/ (accessed May 4, 2022).

40. Solar Geoengineering Non-Use Agreement. https://www.solargeoeng.org/ (accessed May 4, 2022).

41. Wagenknecht, Nils, Andreas Mertens, Olufemi Omitaomu, and Adedeji Badiru (2022), "Legal and Business Aspects of COP26 Agreements," Research Exchange Research Report, Graduate School of Engineering and Management, Air Force Institute of Technology, Wright-Patterson Air Force Base, Dayton, Ohio, USA.

7 Communication, Cooperation, and Coordination for Climate Response

TRIPLE C FRAMEWORK FOR ENVIRONMENTAL RESPONSE

Ultimately, getting things done, in the context of environmental response and sustainability, requires the involvement of many participants. Such participation requires communication, cooperation, and coordination. This chapter introduces the Triple C principle of project execution based on stages of communication, cooperation, and coordination (Badiru, 2008). As presented by the Chinese Proverb below, involvement of every team member is critical for overall success of a project.

Tell me, and I forget;
Show me, and I remember;
Involve me, and I understand.
—**Chinese Proverb**

Triple C model facilitates better understanding and involvement based on foundational communication. The Triple C approach elucidates the integrated involvement of communication, cooperation, and coordination. Communication is the foundation for cooperation, which in turn is the foundation for coordination. Communication leads to cooperation, which leads to coordination, which leads to project harmony, which leads to project success.

The primary lesson of the Triple C model is not to take cooperation for granted. It must be pursued, solicited, and secured explicitly. The process of securing cooperation requires structured communication upfront. It is only after cooperation is in effect that all project efforts can be coordinated.

The Triple C model has been used effectively in practice to enhance project performance because most project problems can be traced to initial communication problems. The Triple C approach works because it is very simple; simple to understand, and simple to implement. The simplicity comes from the fact that most of the required elements of the approach are already being done within every organization, albeit in a non-structured manner. The Triple C model puts the existing processes into a structural approach to communication, cooperation, and coordination.

DOI: 10.1201/9781003279051-7

The idea for the Triple C model originated from a complex facility redesign project (Badiru et al., 1993) conducted for Tinker Air Force Base (TAFB) in Oklahoma City by the School of Industrial Engineering, University of Oklahoma from 1985 through 1989. The project was a part of a reconstruction project following a disastrous fire that occurred in the base's repair/production facility in November 1984. The urgency, complexity, scope ambiguity, confusion, and disjointed directions that existed in the early days of the reconstruction effort led to the need to develop a structured approach to communication, cooperation, and coordination of the various work elements. In spite of the high pressure timing of the project, the author called a Time-Out-Of-time (TOOT) so that a process could be developed for project communication, leading to personnel cooperation; and eventually facilitating task coordination. The investment of TOOT time resulted in a remarkable resurgence of cooperation where none existed at the beginning of the project. Encouraged by the intrinsic occurrence of cooperation, the process was further enhanced and formalized as the Triple C approach to the project's success. The approach was credited with the overall success of the project. The qualitative approach of Triple C complemented the technical approaches used on the project to facilitate harmonious execution of tasks. Many projects fail when the stakeholders get too wrapped up into the technical requirements at the expense of qualitative requirements. Other elements of "C," such as Collaboration, Commitment, and Correlation, are embedded in the Triple C structure. Of course, the constraints of time, cost, and performance must be overcome all along the way.

Organizations thrive by investing in three primary resources: **People** who do the work; the **Tools** that the people use to do the work; and the **Process** that governs the work that the people do. Of the three, investing in people is the easiest thing an organization can do and we should do it whenever we have an opportunity. The Triple C approach incorporates the qualitative (human) aspects of a project into overall project requirements.

The Triple C model was first used in 1985 and subsequently introduced in print in 1987 (Badiru, 1987). The project scenario that led to the development of the Triple Model was later documented in Badiru et al. (1993). The model is an effective project planning and control tool. The model states that project management can be enhanced by implementing it within the integrated functions summarized below:

- Communication
- Cooperation
- Coordination

The model facilitates a systematic approach to project planning, organizing, scheduling, and control. The Triple C model is distinguished from the 3C approach commonly used in military operations. The military approach emphasizes personnel management in the hierarchy of command, control, and communication. This places communication as the last function. The Triple C, by contrast, suggests communication as the first and foremost function. The Triple C model can be implemented for project planning, scheduling and control purposes.

Figure 7.1 shows the application of Triple C for project planning, scheduling, and control within the confines of the Triple Constraints of cost, schedule, and performance.

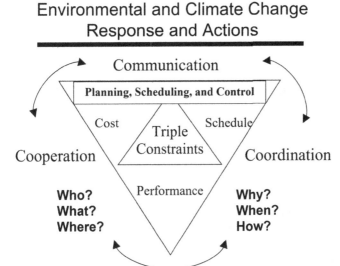

FIGURE 7.1   Triple C for Communication, Cooperation, and Coordination.

Each of these three primary functions of project management requires effective communication, sustainable cooperation, and adaptive coordination.

Triple C illustrates the basic questions of what, who, why, how, where, and when. It highlights what must be done and when. It can also help to identify the resources (personnel, equipment, facilities, etc.) required for each effort. It points out important questions such as

- Does each project participant know what the objective is?
- Does each participant know his or her role in achieving the objective?
- What obstacles may prevent a participant from playing his or her role effectively?

Triple C can mitigate disparity between idea and practice because it explicitly solicits information about the critical aspects of a project in terms of the following queries:

TYPES OF COMMUNICATION

- Verbal
- Written
- Body language
- Visual tools (e.g., graphical tools)
- Sensual (use of all five senses: sight, smell, touch, taste, hearing; olfactory, tactile, auditory)
- Simplex (unidirectional)
- Half-duplex (bi-directional with time lag)

- Full-duplex (real-time dialogue)
- One-on-one
- One-to-many
- Many-to-one

TYPES OF COOPERATION

- Proximity
- Functional
- Professional
- Social
- Romantic
- Power influence
- Authority influence
- Hierarchical
- Lateral
- Cooperation by intimidation
- Cooperation by enticement

TYPES OF COORDINATION

- Teaming
- Delegation
- Supervision
- Partnership
- Token-passing
- Baton hand-off

TRIPLE C QUESTIONS

Questioning is the best approach to getting information for effective project management. Everything should be questioned. By upfront questions, we can preempt and avert project problems later on. Typical questions to ask under the Triple C approach are:

- What is the purpose of the project?
- Who is in charge of the project?
- Why is the project needed?
- Where is the project located?
- When will the project be carried out?
- How will the project contribute to increased opportunities for the organization?
- What is the project designed to achieve?
- How will the project affect different groups of people within the organization?
- What will be the project approach or methodology?
- What other groups or organizations will be involved (if any)?

- What will happen at the end of the project?
- How will the project be tracked, monitored, evaluated, and reported?
- What resources are required?
- What are the associated costs of the required resources?
- How do the project objectives fit the goal of the organization?
- What respective contribution is expected from each participant?
- What level of cooperation is expected from each group?
- Where is the coordinating point for the project?

The key to getting everyone on board with a project is to ensure that task objectives are clear and comply with the principle of **SMART**, as outlined below:

Specific: Task objective must be specific.
Measurable: Task objective must be measurable.
Aligned: Task objective must be achievable and aligned with overall project goal.
Realistic: Task objective must be realistic and relevant to the organization.
Timed: Task objective must have a time basis.

If a task has the above intrinsic characteristics, then the function of communicating the task will more likely lead to personnel cooperation.

COMMUNICATION

Communication makes working together possible. The communication function of project management involves making all those concerned become aware of project requirements and progress. Those who will be affected by the project directly or indirectly, as direct participants or as beneficiaries, should be informed as appropriate regarding the following:

- Scope of the project
- Personnel contribution required
- Expected cost and merits of the project
- Project organization and implementation plan
- Potential adverse effects if the project should fail
- Alternatives, if any, for achieving the project goal
- Potential direct and indirect benefits of the project

The communication channel must be kept open throughout the project life cycle. In addition to internal communication, appropriate external sources should also be consulted. The project manager must

- Exude commitment to the project
- Utilize the communication responsibility matrix
- Facilitate multi-channel communication interfaces

- Identify internal and external communication needs
- Resolve organizational and communication hierarchies
- Encourage both formal and informal communication links

When clear communication is maintained between management and employees and among peers, many project problems can be averted. Project communication may be carried out in one or more of the following formats:

- One-to-many
- One-to-one
- Many-to-one
- Written and formal
- Written and informal
- Oral and formal
- Oral and informal
- Nonverbal gestures

Good communication is affected when what is implied is perceived as intended. Effective communications are vital to the success of any project. Despite the awareness that proper communications form the blueprint for project success, many organizations still fail in their communications functions. The study of communication is complex. Factors that influence the effectiveness of communication within a project organization structure include the following.

1. **Personal perception.** Each person perceives events on the basis of personal psychological, social, cultural, and experimental background. As a result, no two people can interpret a given event the same way. The nature of events is not always the critical aspect of a problem situation. Rather, the problem is often the different perceptions of the different people involved.
2. **Psychological profile.** The psychological makeup of each person determines personal reactions to events or words. Thus, individual needs and level of thinking will dictate how a message is interpreted.
3. **Social Environment.** Communication problems sometimes arise because people have been conditioned by their prevailing social environment to interpret certain things in unique ways. Vocabulary, idioms, organizational status, social stereotypes, and economic situation are among the social factors that can thwart effective communication.
4. **Cultural background.** Cultural differences are among the most pervasive barriers to project communications, especially in today's multinational organizations. Language and cultural idiosyncrasies often determine how communication is approached and interpreted.
5. **Semantic and syntactic factors.** Semantic and syntactic barriers to communications usually occur in written documents. Semantic factors are those that relate to the intrinsic knowledge of the subject of the communication. Syntactic factors are those that relate to the form in which the

communication is presented. The problems created by these factors become acute in situations where response, feedback, or reaction to the communication cannot be observed.

6. **Organizational structure.** Frequently, the organization structure in which a project is conducted has a direct influence on the flow of information and, consequently, on the effectiveness of communication. Organization hierarchy may determine how different personnel levels perceive a given communication.

7. **Communication media.** The method of transmitting a message may also affect the value ascribed to the message and, consequently, how it is interpreted or used. The common barriers to project communications are

- Inattentiveness
- Lack of organization
- Outstanding grudges
- Preconceived notions
- Ambiguous presentation
- Emotions and sentiments
- Lack of communication feedback
- Sloppy and unprofessional presentation
- Lack of confidence in the communicator
- Lack of confidence by the communicator
- Low credibility of communicator
- Unnecessary technical jargon
- Too many people involved
- Untimely communication
- Arrogance or imposition
- Lack of focus.

Some suggestions on improving the effectiveness of communication are presented next. The recommendations may be implemented as appropriate for any of the forms of communications listed earlier. The recommendations are for both the communicator and the audience.

1. Never assume that the integrity of the information sent will be preserved as the information passes through several communication channels. Information is generally filtered, condensed, or expanded by the receivers before relaying it to the next destination. When preparing a communication that needs to pass through several organization structures, one safeguard is to compose the original information in a concise form to minimize the need for recomposition of the project structure.

2. Give the audience a central role in the discussion. A leading role can help make a person feel a part of the project effort and responsible for the projects' success. He or she can then have a more constructive view of project communication.

3. Do homework and think through the intended accomplishment of the communication. This helps eliminate trivial and inconsequential communication efforts.

4. Carefully plan the organization of the ideas embodied in the communication. Use indexing or points of reference whenever possible. Grouping ideas into related chunks of information can be particularly effective. Present the short messages first. Short messages help create focus, maintain interest and prepare the mind for the longer messages to follow.

5. Highlight why the communication is of interest and how it is intended to be used. Full attention should be given to the content of the message with regard to the prevailing project situation.

6. Elicit the support of those around you by integrating their ideas into the communication. The more people feel they have contributed to the issue, the more expeditious they are in soliciting the cooperation of others. The effect of the multiplicative rule can quickly garner support for the communication purpose.

7. Be responsive to the feelings of others. It takes two to communicate. Anticipate and appreciate the reactions of members of the audience. Recognize their operational circumstances and present your message in a form they can relate to.

8. Accept constructive criticism. Nobody is infallible. Use criticism as a springboard to higher communication performance.

9. Exhibit interest in the issue in order to arouse the interest of your audience. Avoid delivering your messages as a matter of a routine organizational requirement.

10. Obtain and furnish feedback promptly. Clarify vague points with examples.

11. Communicate at the appropriate time, at the right place, to the right people.

12. Reinforce words with positive action. Never promise what cannot be delivered. Value your credibility.

13. Maintain eye contact in oral communication and read the facial expressions of your audience to obtain real-time feedback.

14. Concentrate on listening as much as speaking. Evaluate both the implicit and explicit meanings of statements.

15. Document communication transactions for future references.

16. Avoid asking questions that can be answered yes or no. Use relevant questions to focus the attention of the audience. Use questions that make people reflect upon their words, such as, "How do you think this will work?" compared to "Do you this will work?"

17. Avoid patronizing the audience. Respect their judgment and knowledge.

18. Speak and write in a controlled tempo. Avoid emotionally charged voice inflections.

19. Create an atmosphere for formal and informal exchange of ideas.

20. Summarize the objectives of the communication and how they will be achieved.

Within the framework of Triple C, a communication responsibility matrix contains the linking of sources of communication and targets of communication. Cells within the matrix indicate the subject of the desired communication. There should be at least one filled cell in each row and each column of the matrix. This assures that each individual of a department has at least one communication source or target associated with him or her. With a communication responsibility matrix, a clear understanding of what needs to be communicated to whom can be developed. Communication in a project environment can take any of several forms. The specific needs of a project may dictate the most appropriate mode. Three popular computer communication modes are discussed next in the context of communicating data and information for project management.

Simplex communication. This is a unidirectional communication arrangement in which one project entity initiates communication to another entity or individual within the project environment. The entity addressed in the communication does not have mechanism or capability for responding to the communication. An extreme example of this is a one-way, top-down communication from top management to the project personnel. In this case, the personnel have no communication access or input to top management. A budget-related example is a case where top management allocates budget to a project without requesting and reviewing the actual needs of the project. Simplex communication is common in authoritarian organizations.

Half-duplex communication. This is a bi-directional communication arrangement whereby one project entity can communicate with another entity and receive a response within a certain time lag. Both entities can communicate with each other but not at the same time. An example of half-duplex communication is a project organization that permits communication with top management without a direct meeting. Each communicator must wait for a response from the target of the communication. Request and allocation without a budget meeting is another example of half-duplex data communication in project management.

Full-duplex communication. This involves a communication arrangement that permits a dialogue between the communicating entities. Both individuals and entities can communicate with each other at the same time or face-to-face. As long as there is no clash of words, this appears to be the most receptive communication mode. It allows participative project planning in which each project personnel has an opportunity to contribute to the planning process.

Each member of a project team needs to recognize the nature of the prevailing communication mode in the project. Management must evaluate the prevailing communication structure and attempt to modify it if necessary to enhance project functions. An evaluation of who is to communicate with whom about what may help improve the project data/information communication process. A communication matrix may include notations about the desired modes of communication between individuals and groups in the project environment.

COOPERATION

The cooperation of the project personnel must be explicitly elicited. Merely voicing consent for a project is not enough assurance of full cooperation. The participants and beneficiaries of the project must be convinced of the merits of the project. Some of the factors that influence cooperation in a project environment include personnel requirements, resource requirements, budget limitations, past experiences, conflicting priorities, and lack of uniform organizational support. A structured approach to seeking cooperation should clarify the following:

- Cooperative efforts required
- Precedents for future projects
- Implication of lack of cooperation
- Criticality of cooperation to project success
- Organizational impact of cooperation
- Time frame involved in the project
- Rewards of good cooperation

Cooperation is a basic virtue of human interaction. More projects fail due to a lack of cooperation and commitment than any other project factors. To secure and retain the cooperation of project participants, you must elicit a positive first reaction to the project. The most positive aspects of a project should be the first items of project communication. For project management, there are different types of cooperation that should be understood.

Functional cooperation. This is cooperation induced by the nature of the functional relationship between two groups. The two groups may be required to perform related functions that can only be accomplished through mutual cooperation.

Social cooperation. This is the type of cooperation effected by the social relationship between two groups. The prevailing social relationship motivates cooperation that may be useful in getting project work done.

Legal cooperation. Legal cooperation is the type of cooperation that is imposed through some authoritative requirement. In this case, the participants may have no choice other than to cooperate.

Administrative cooperation. This is cooperation brought on by administrative requirements that make it imperative that two groups work together on a common goal.

Associative cooperation. This type of cooperation may also be referred to as collegiality. The level of cooperation is determined by the association that exists between two groups.

Proximity cooperation. Cooperation due to the fact that two groups are geographically close is referred to as proximity cooperation. Being close makes it imperative that the two groups work together.

Dependency cooperation. This is cooperation caused by the fact that one group depends on another group for some important aspect. Such dependency is usually of a mutual two-way nature. One group depends on the other for one thing while the latter group depends on the former for some other thing.

Imposed cooperation. In this type of cooperation, external agents must be employed to induce cooperation between two groups. This is applicable for cases where the two groups have no natural reason to cooperate. This is where the approaches presented earlier for seeking cooperation can became very useful.

Lateral cooperation. Lateral cooperation involves cooperation with peers and immediate associates. Lateral cooperation is often easy to achieve because existing lateral relationships create an environment that is conducive for project cooperation.

Vertical cooperation. Vertical or hierarchical cooperation refers to cooperation that is implied by the hierarchical structure of the project. For example, subordinates are expected to cooperate with their vertical superiors.

Whichever type of cooperation is available in a project environment, the cooperative forces should be channeled toward achieving project goals. Documentation of the prevailing level of cooperation is useful for winning further support for a project. Clarification of project priorities will facilitate personnel cooperation. Relative priorities of multiple projects should be specified so that a priority to one group is also a priority to all groups within the organization. Some guidelines for securing cooperation for most projects are

- Establish achievable goals for the project.
- Clearly outline the individual commitments required.
- Integrate project priorities with existing priorities.
- Eliminate the fear of job loss due to industrialization.
- Anticipate and eliminate potential sources of conflict.
- Use an open-door policy to address project grievances.
- Remove skepticism by documenting the merits of the project.

Commitment. Cooperation must be supported with commitment. To cooperate is to support the ideas of a project. To commit is to willingly and actively participate in project efforts again and again through the thick and thin of the project. Provision of resources is one way that management can express commitment to a project.

COORDINATION

After the communication and cooperation functions have successfully been initiated, the efforts of the project personnel must be coordinated. Coordination facilitates harmonious organization of project efforts. The construction of a responsibility chart can be very helpful at this stage. A responsibility chart is a matrix consisting of columns of individual or functional departments and rows of required actions. Cells within the matrix are filled with relationship codes that indicate who is responsible for what. Table 7.1 illustrates an example of a responsibility matrix for the planning for a seminar program. The matrix helps avoid neglecting crucial communication requirements and obligations. It can help resolve questions such as

- Who is to do what?
- How long will it take?
- Who is to inform whom of what?
- Whose approval is needed for what?
- Who is responsible for which results?
- What personnel interfaces are required?
- What support is needed from whom and when?

TABLE 7.1
Example of Responsibility Matrix for Project Coordination

TASKS	Person Responsible				Status of Task			
	Staff A	Staff B	Staff C	Mgr	31-Jan	15-Feb	28-Mar	21-Apr
Brainstorming Meeting	R	R	R	R	D			
Identify Speakers	I	R		R				
Select Seminar Location	R	R	R			O		
Select Banquet Location	R	R				O		
Prepare Publicity Materials		C	R	I	O	O	D	
Draft Brochures		C	R				D	D
Develop Schedule			R		L	L	L	
Arrange for Visual Aids			R			L	L	
Coordinate Activities			R			L	L	
Periodic Review of Tasks	R	R	R	S				D
Monitor Progress of Program	C	R	R			O	L	
Review Program Progress	R				O	O	L	L
Closing Arrangements	R							L
Post-Program Review and Evaluation	R	R	R	R		D	D	

Responsibility Codes:
R = Responsible
I = Inform
S = Support
C = Consult
Task Codes:
D = Done
O = On Track
L = Late

CONFLICT RESOLUTION USING TRIPLE C APPROACH

Conflicts can and do develop in any work environment. Conflicts, whether intended or inadvertent, prevents an organization from getting the most out of the workforce. When implemented as an integrated process, the Triple C model can help avoid conflicts in a project. When conflicts do develop, it can help in resolving the conflicts. The key to conflict resolution is open and direct communication, mutual cooperation, and sustainable coordination. Several sources of conflicts can exist in a projects. Some of these are discussed below.

Schedule conflict. Conflicts can develop because of improper timing or sequencing of project tasks. This is particularly common in large multiple projects. Procrastination can lead to having too much to do at once, thereby creating a clash of project functions and discord among project team members. Inaccurate estimates of time requirements may lead to infeasible activity schedules. Project coordination can help avoid schedule conflicts.

Cost conflict. Project cost may not be generally acceptable to the clients of a project. This will lead to project conflict. Even if the initial cost of the project is acceptable, a lack of cost control during implementation can lead to conflicts. Poor budget allocation approaches and the lack of a financial feasibility study will cause cost conflicts later on in a project. Communication and coordination can help prevent most of the adverse effects of cost conflicts.

Performance conflict. If clear performance requirements are not established, performance conflicts will develop. Lack of clearly defined performance standards can lead each person to evaluate his or her own performance based on personal value judgments. In order to uniformly evaluate quality of work and monitor project progress, performance standards should be established by using the Triple C approach.

Management conflict. There must be a two-way alliance between management and the project team. The views of management should be understood by the team. The views of the team should be appreciated by management. If this does not happen, management conflicts will develop. A lack of a two-way interaction can lead to strikes and industrial actions, which can be detrimental to project objectives. The Triple C approach can help create a conducive dialogue environment between management and the project team.

Technical conflict. If the technical basis of a project is not sound, technical conflict will develop. New industrial projects are particularly prone to technical conflicts because of their significant dependence on technology. Lack of a comprehensive technical feasibility study will lead to technical conflicts. Performance requirements and systems specifications can be integrated through the Triple C approach to avoid technical conflicts.

Priority conflict. Priority conflicts can develop if project objectives are not defined properly and applied uniformly across a project. Lack of a direct project definition can lead each project member to define his or her own goals which may be in conflict with the intended goal of a project. Lack of consistency of the project mission is another potential source of priority conflicts. Over-assignment of responsibilities with no guidelines for relative significance levels can also lead to priority conflicts. Communication can help defuse priority conflict.

Resource conflict. Resource allocation problems are a major source of conflict in project management. Competition for resources, including personnel, tools, hardware, software, and so on, can lead to disruptive clashes among project members. The Triple C approach can help secure resource cooperation.

Power conflict. Project politics lead to a power play which can adversely affect the progress of a project. Project authority and project power should be clearly delineated. Project authority is the control that a person has by virtue of his or her functional post. Project power relates to the clout and influence, which a person can exercise due to connections within the administrative structure. People with popular personalities can often wield a lot of project power in spite of low or nonexistent project authority. The Triple C model can facilitate a positive marriage of project authority and power to the benefit of project goals. This will help define clear leadership for a project.

Personality conflict. Personality conflict is a common problem in projects involving a large group of people. The larger the project, the larger the size of the management team needed to keep things running. Unfortunately, the larger management team creates an opportunity for personality conflicts. Communication and cooperation can help defuse personality conflicts. In summary, conflict resolution through Triple C can be achieved by observing the following guidelines:

1. Confront the conflict and identify the underlying causes.
2. Be cooperative and receptive to negotiation as a mechanism for resolving conflicts.
3. Distinguish between proactive, inactive, and reactive behaviors in a conflict situation.
4. Use communication to defuse internal strife and competition
5. Recognize that short-term compromise can lead to long-term gains.
6. Use coordination to work toward a unified goal.
7. Use communication and cooperation to turn a competitor into a collaborator.

It is the little and often-neglected aspects of a project that lead to project failures. Several factors may constrain the project implementation. All the relevant factors can be evaluated under the Triple C model right from the project-initiation stage.

APPLICATION OF TRIPLE C TO ENVIRONMENTAL PROJECTS

Having now understood the intrinsic elements of Triple C, we can see how and where it could be applicable to environmental project management. Communication explains project scope and requirements through the stages of planning, organizing, scheduling, and control. Cooperation is required to get human resource buy-in and stakeholder endorsement across all facets of planning, organizing, scheduling, and control. Coordination facilitates adaptive interfaces over all the elements of planning, organizing, scheduling, and control. The Triple C model should be implemented as an iterative loop process that moves a project through the communication, cooperation, and coordination functions.

DMAIC AND TRIPLE C

Many organizations now explore Six Sigma DMAIC (Define, Measure, Analyze, Improve and Control) methodology and associated tools to achieve better project performance. Six Sigma means six standard deviations from a statistical performance average. The Six Sigma approach allows for no more than 3.4 defects per million parts in manufactured goods or 3.4 mistakes per million activities in a service operation. To explain the effect of the Six Sigma approach, consider a process that is 99% perfect. That process will produce 10,000 defects per million parts. With Six Sigma, the process will need to be 99.99966% perfect in order to produce only 3.4 defects per million. Thus, Six Sigma is an approach that moves a process toward perfection. Six Sigma, in effect, reduces variability among products produced by the same process. By contrast, the Lean approach is designed to reduce/eliminate waste in the production process.

Six Sigma provides a roadmap for the five major steps of DMAIC (Define, Measure, Analyze, Improve and Control), which are applicable to the planning and control steps of project management. We cannot improve what we cannot measure. Triple C provides a sustainable approach to obtaining cooperation and coordination for DMAIC during improvement efforts. DMAIC requires project documentation and reporting, which coincide with project control requirements.

CONCLUSIONS

This chapter has presented a general introduction to the Triple C approach, focusing on Communication, Cooperation, and Coordination. A summary of lessons to be inferred from a Triple C approach are:

- Use proactive planning to initiate project functions.
- Use preemptive planning to avoid project pitfalls.
- Use meetings strategically. Meeting is not *work*. Meeting should be done to facilitate work.
- Use project assessment to properly frame the problem, adequately define the requirements, continually ask the right questions, cautiously analyze risks, and effectively scope the project.
- Be bold to terminate a project when termination is the right course of action. Every project needs an exit plan. In some cases, there is victory in capitulation.

The applicability and sustainability of the Triple C approach is summarized below:

1. For effective communication, create good communication channels.
2. For enduring cooperation, establish partnership arrangements.
3. For steady coordination, use a workable organization structure.

Further details on Triple C can be found in Badiru (2008).

REFERENCES

Badiru, Adedeji B., (1987), "Communication, Cooperation, Coordination: The Triple C of Project Management," *Proceedings of 1987 IIE Spring Conference*, Washington, DC, May 1987, pp. 401–404.

Badiru, Adedeji B., (2008), *Triple C Model of Project Management: Communication, Cooperation, and Coordination*, Taylor & Francis Group/CRC Press, Boca Raton, FL.

Badiru, Adedeji B., B. L. Foote, L. Leemis, A. Ravindran, and L. Williams (1993), "Recovering from a Crisis at Tinker Air Force Base," *PM Network*, Vol. 7, No. 2, Feb. 1993, pp. 10–23.

8 Recycling of End-of-Life Vehicles

ADAPTED WITH OPEN-ACCESS PERMISSION FROM

Al-Quradaghi, Shimaa, Quipeng P. Zheng, and Ali Eklamel (2020), Generalized Framework for the Design of Eco-Industrial Parks: Case Study of End-of-Life Vehicles, *Sustainability* 2020, *12*, 6612; doi:10.3390su12166612

BACKGROUND

Al-Quradaghi et al. (2020) present an excellent example of how industrial development pursuits can take cognisance of the global push for sustainability. This chapter presents a condensed synopsis of the case study project published by the authors. Without a loss of generality and due to space limitation, only the key conceptual framework of the case study is presented here without the figures and artwork of the study report. Interested readers are directed to Al-Quradaghi et al. (2020) for the full details.

Eco-industrial parks (EIPs) are promoting a shift from the traditional linear model to the circular model, where industrial symbiosis plays an important role in encouraging the exchange of materials, energy, and waste. This chapter proposes a generalized framework to design eco-industrial parks (EIPs), and illustrates it with regard to the end-of-life vehicle problem (ELV). An eco-industrial park for end-of-life vehicles (EIP-4-ELVs) creates synergy in the network that leverages waste reduction and makes efficient use of resources. The performance of the proposed framework is investigated, along with the interactions between nodes. The proposed framework consists of five steps: (1) finding motivation for EIP, (2) identifying all entities with industrial symbiosis, (3) pinpointing the anchor entity, (4) determining industrial symbiosis between at least three entities and two exchange flows, and (5) defining exchange-flow types. The two last steps are connected by a feedback loop to allow future exchange flows. The proposed framework serves as a guideline for decision-makers during the first stages of developing EIPs. Furthermore, the framework can be linked to car-design software to predict the recyclability of vehicle components and aid in manufacturing vehicles optimized for recycling.

INTRODUCTION

An industrial ecosystem optimizes the consumption of energy and materials, and minimizes the generation of waste [1, 7, 52]. The study of industrial systems that operate like natural ecosystems is called industrial ecology, in which the natural ecosystem generates waste from one organism to be the resource for another [2]. Similar to the natural ecosystem, the industrial ecosystem capitalizes on the exchanges of one firm's waste to be another firm's resource.

Industrial symbiosis is a subset of industrial ecology and has a particular focus on material and energy exchanges [3]. Eco-industrial parks (EIPs) develop when industrial symbiosis occurs between firms; the interactions include exchanges of material and energy.

EIPs promote a shift from the traditional linear model to the circular model. They are considered a community of businesses that reduce the global impact by sharing resources like materials, energy, and water to reduce waste and pollution, and increase economic gains [4]. The interactions in the community improve the environmental performance of the industrial network. EIP is promoting a shift from the traditional linear model of "raw material to industry to waste" to a closed-loop model of "raw material to industry A to waste to raw material to industry B." Considering the global strategy, cities are considered at the macro level, and single industries at the micro level, and, in between, there is the meso level where EIPs exist. EIPs exchange not only waste heat, steam, bio-waste, and industrial waste but also knowledge, material, and energy. The circular process that is built within EIP connects the entities and results with a minimum impact on the environment. Hence, solving environmental problems using an EIPs approach is one effective strategy of waste management that limits pollution impacts on the environment. In this way, EIPs can be implemented within industrial districts to encourage eco-design practices and the transition to circular business models that are more sustainable in nature. This will also improve the efficiencies of the existing industrial entities in given districts and thus enables them to improve their competitiveness on a global scale. The first known EIP was in Kalundborg, Denmark; industrial symbiosis gradually evolved over 20 years [5]. Today, many countries have eco-industrial park projects: Argentina, Austria, Brazil, Canada, China, Denmark, Finland, France, Germany, Italy, Netherlands, Norway, Singapore, South Korea, Spain, Sweden, Switzerland, the UK, the USA, and others [3, 4, 6–11]. Other works vis-à-vis EIPs covered multidisciplinary areas, for example, optimization [12], life-cycle assessments [13], policy implementation [14], social networks [15], and topology [4]. Furthermore, several studies in the literature proposed frameworks for EIPs. For example, in December 2017, the United Nations Industrial Development Organization (UNIDO), in a joint initiative with the German Development Co-Operation (GIZ) and the World Bank Group (WBG), published a document that presents an international framework for eco-industrial parks [16]. The International Framework for Eco-Industrial Parks focuses on four performance categories: park-management, environmental, social, and economic (see Al-Quradaghi et al., 2020). Each category consists of prerequisites and performance requirements that can be measured. All prerequisites and performance requirements must be met for something to be considered an EIP. The report guides decision-makers on the important components to achieve maximal benefits economically, environmentally, and socially. However, as indicated in the report, the International Framework for EIPs provides only strategic details for EIP requirements and does not translate them to existing EIPs.

The International Framework for Eco-Industrial Parks created by UNIDO, GIZ, and WBG is not the only initiative to provide essential elements for forming EIPs. Several studies in the literature proposed frameworks for EIPs from different views. There are two common themes for these frameworks, general and special cases. While researchers like [17–25] outlined a general framework for EIPs, other researchers, like [7, 26, 27] proposed more specific frameworks for solving a special-case issue. The reviewed articles are presented in the next sections in chronological order in each theme.

Haskins (2007) proposed a general framework for eco-industrial park development named iFACE. The acronym breaks down as follows: i—identify stakeholders and their needs; F—frame the problem(s); A—alternatives identification and study the options; C—choose and implement a course of action; and E—evaluate continuously. The author expressed the framework as a combination of system engineering, industrial ecology, organizational dynamics, logistics, and supply-chain theories [17].

Sopha et al. (2009) presented a more extended framework for creating industrial-symbiosis modeling. The framework consists of two parts: (1) a system-engineering (SE) process; and (2) methods. The SE process consists of six steps: needs identification, defining requirements, specifying performances, analyzing, designing and improving, and implementation. The methods component lists different enabling technologies for each SE process step. Interviewing was proposed for Steps 1–3, brainstorming for Step 1, literature study for Steps 2–6, survey for Steps 2–4, field study for Steps 1–3, and workshop for Steps 3–6. The authors applied the framework on the case of an industrial site in Mongstad, Norway, to increase industrial symbiosis [18].

Boons et al. (2011) proposed a conceptual framework for analyzing the dynamics of industrial symbiosis. The framework has a set of conditions that are referred to as "antecedents" that affects another set of "mechanisms." The mechanisms elaborate on two levels: (1) societal and (2) regional industrial systems. The application of the mechanisms led to outcomes that were reflected on the ecological system and social networks. The authors concluded that the framework helped to build a theoretical understanding of the dynamics of industrial symbiosis [19].

Romero and Ruiz (2013) proposed a nested-system framework for modeling EIP operations. The framework describes the relationship between industrial systems and the environment. The main building blocks for the nested framework are economic, social, and natural systems. In applying the framework, five key properties were taken into consideration: (1) functionality; (2) theoretical knowledge; (3) adaptability; (4) reliability; and (5) life span. The authors supported the framework by merging complex-adaptive-system theory, industrial ecology, and the analysis of existing EIPs [24].

Francois Dumoulin et al. (2016) proposed an environmental-assessment framework for facilitated regional industrial symbiosis. The framework helps to identify all environmental impacts in facilitated regional industrial symbiosis. The framework was divided into two main sections: (1) logical basis, where key elements of environment observation are identified; and (2) method, where three steps are performed, namely, identifying environmental impact, designing indicators, and assessing the environment. The authors applied the framework on a case in Réunion, a French territory in the Indian Ocean that had the potential for industrial symbiosis [20].

Kuznetsova et al. (2016) discussed the challenges faced by EIPs and proposed an optimization framework for EIP topology and operation. The framework consisted of two stages: (1) optimization of EIPs' topology and (2) operation. Each stage includes several steps. The framework considers uncertainties in EIP and provides appropriate predictions. The authors detailed uncertainties and risks that should be taken into consideration at the design stage [28].

Andiappan et al. (2016) proposed an optimization-based framework for coalitions in EIP. The framework starts with defining co-operative plants in the intended EIP. Then, the framework continues identifying interactions between plants, uses mathematical models to calculate the symbiosis costs (cost of sharing exchanges between plants), and ends up with the eco-industrial park configuration. The authors used mathematical models to calculate the economic correlations of cost and savings allocation, and performed stability analysis for each entity. The framework was applied to a palm-oil EIP in Malaysia. The results showed an increase in the savings for industries in the EIP [22].

Konstantinova, Johannes, and Vejrum (2019) discussed the importance of trust between stakeholders in industrial-symbiosis initiatives. They developed a conceptual industrial-symbiosis trust framework to facilitate gaining trust between partners. The framework illustrates three notions of trust (ability, integrity, and benevolence) through three trust bases: (1) calculus-, (2) knowledge-, and (3) identification-based trusts. In their paper, the authors answered the research question "How can firms develop trust in the context of industrial symbioses investment?" through proposing the framework that merges industrial symbiosis and management techniques [21].

Tao et al. (2019) proposed a three-dimensional framework for studying the influence of policy on industrial symbiosis from the firm's perspective. The three dimensions are industrial-symbiosis (1) fostering models, (2) implementation stages, and (3) policy instruments. The framework was demonstrated on a horizontal axis that presented ten executive policy instruments, the vertical axis presented five stages of industrial-symbiosis implementation, and the depth axis presented four models of industrial-symbiosis fostering. The authors applied the framework on two existing EIPs, one in the United Kingdom and the other in China [25].

For special case frameworks, Behera et al. (2012) presented the existing Research and Development into Business framework that was developed by the Ulsan EIP Center in South Korea. The framework consists of three main steps: (1) exploring new networks; (2) feasibility study; and (3) commercialization. Each step leads to another step following certain criteria within the framework. The aim of the framework was to design industrial symbiosis between EIP stockholders. As an example, the authors presented the Ulsan industrial symbiosis and explained how it had been developed to reach forty instances of symbiosis, some of which having been designed [7].

Liu and Côté (2017) presented a framework for incorporating ecosystem services into China's EIPs. The framework combines policies, governance, techniques, technologies, key actors, and support organizations to build the industrial symbiosis. In the framework, two main components are the core: Component I, with elements of policies, governance, technologies, and business development; and Component II, with elements of key actors and support organizations. The framework suggested integration between the two components to result in an EIP. The framework was designed to solve China's environmental issues by proposing a circular economy through encouraging eco-industrial park development. The authors suggested that the framework could provide guidance for other EIPs around the world [26].

Gopinath et al. (2018) presented a material-flow framework for the sugar industry on the basis of an extensive literature review. The authors reviewed the literature to find all characteristics of the sugar industry to identify the optimal route that resulted in minimizing waste. The framework details the material flow for the sugar industry and points to several waste types that can be reused by other industries. The paper shows the importance of synergy between different types of sectors [27].

The above sections provided an overview of some available frameworks for EIPs in the literature. The discussions outlined the efforts by scholars with different backgrounds to propose ways and methods for designing EIPs. The review elucidates that although considerable efforts have been done in the literature, essential research regarding applying the approach to different types of systems such as ELVs is still required. Furthermore, the literature revealed the complexity of the available frameworks. In the early stage of designing EIPs, decision-makers need a simple, clear, and strategic framework to follow and depend upon. After having a solid base about the foundation of the elements in EIPs with a general clear framework, decision-makers and the EIP team might follow other comprehensive frameworks that pertain to specific needs, such as optimizing industrial-symbiosis flow, evaluating the firms' trust, proposing policies, and other areas of concern. There exists no generalized framework that compiles precise foundations and simple steps at the same time. Hence, there is a need to investigate and construct a generalized and simple-to-follow framework for the design of eco-industrial parks. This chapter bridges this gap in the literature. A simple and yet comprehensive framework is proposed in this chapter. The framework consists of five steps: (1) finding motivation for EIP; (2) identifying all entities with industrial symbiosis; (3) pinpointing the anchor entity; (4) determining industrial symbiosis between at least three entities and two exchange flows; and (5) defining exchange-flow types. Steps (4) and (5) are connected by a feedback loop, which allows any additional exchange flows in the future. The chapter illustrates the use of the framework on a special case study that involves end-of-life vehicles. Applying the recyclability index proposed by Villalba et al. (2004), the framework can be linked to vehicle-design software to predict the recyclability of different types of components [29].

The remainder of this chapter is organized as follows. The next section presents the proposed framework for efficiently designing eco-industrial parks. Then, a case study is employed to illustrate the use of the framework. The final section outlines the conclusions and future work.

GENERALIZED FRAMEWORK FOR ECO-PARK DESIGN

With the extensive approaches available in the literature, the need for a straightforward framework is rising. The proposed framework in this section illustrates the foundations of designing eco-industrial parks. It provides general step-by-step actions to be followed in the very early stages of designing EIPs. The framework answers fundamental questions that decision-makers need in order to start developing and designing EIPs. In a simple, clear, and step-by-step strategy, the framework lays out necessary actors in the process of designing EIPs. The framework requires primary data that can easily be collected from each entity forming EIPs.

The generalized framework for the design of EIPs consists of five steps (see Al-Quradaghi et al., 2020): (1) finding motivation for EIP [28]; (2) identifying all entities with industrial symbiosis [30]; (3) pinpointing the anchor entity [31]; (4) determining industrial symbiosis between at least three entities and two exchange flows [4]; and (5) defining exchange-flow types [31]. Steps (4) and (5) are connected by a feedback loop to allow additional exchange flows in the future. Used parts should be first put into a used market depending on historical demand. Information on their utilization should be monitored, and a decision to eventually move them to the recycling node and consider them as waste material should be made frequently. This balance between reuse of useful components and recycling them as waste is an important consideration in Step 2 of the proposed methodology.

To create a successful EIP, a clear motivation should be stated, whether environmental, economic, or mixed [30]. From this point forward, identifying entities shapes the overall components of the network. Next, pointing out the main entity in the network plays a great role in outlining the anchor entity that attracts other entities towards it.

To ensure industrial symbiosis in the EIP, the 3–2 heuristic rule [4] should be satisfied. This rule states that "at least three different entities must be involved in exchanging at least two different resources" [4]. Hence, it is necessary to generate a full list of possible exchanges in the network [4, 32–34]. For this reason, the EIP matrix is proposed to list all industrial symbioses in EIPs.

The last step is to identify material-exchange types. A figure presented in Al-Quradaghi et al. (2020) illustrates the different exchange types on the basis of the literature. In her study, to find the taxonomy of different material-exchange types, (Chertow, 2000) [30] classified five types: Type 1: Through waste exchanges via a third party; Type 2: Within facility, firm, or organization; Type 3: Between firms collocated in a defined eco-industrial park; Type 4: Between local firms that are not collocated; and Type 5: Between virtually organized firms across a broader region [4, 30]. However, she highlighted that "Types 3–5 offer approaches that can readily be identified as industrial symbiosis" [30]. For that reason, we considered material-exchange Types 3–5 to be the main types for flows to build EIP. On that basis, we developed an illustrative figure to simplify external-exchange Types 3–5.

The framework has a feedback loop between Steps (4) and (5) that allows for modification and addition whenever there is new industrial symbiosis in the system. The framework results in a connected graph with nodes and arcs. Nodes represent the entities in the network, and arcs represent the exchange flows (see Al-Quradaghi et al., 2020). The framework steps are elaborated upon in the next subsection.

IDENTIFY MOTIVATION FOR SUSTAINABILITY

This step initially ensures the commitment of creating EIPs. The question, "Why create an EIP?" is very important, and, on the basis of the answer, the motivation is clear. The motivation for forming EIP can be related to the three pillars of sustainability: economic, environmental, and social. Some companies exchange resources in order to reduce cost or increase profit (economic pillar). Other companies have industrial symbiosis as a way to reduce greenhouse-gas (GHG) emissions and waste (environmental pillar). To go beyond these two reasons, some companies form EIPs to create more job opportunities to people in the EIP region (social pillar). In fact, all these benefits can be met in EIP.

IDENTIFY ENTITIES

After identifying the motivation behind forming an EIP, the next step is to identify all possible entities that help in achieving that motivation. Entities in the planned EIP should have exchange flows with other entities from which the EIP benefits.

PINPOINT ANCHOR ENTITY

From all entities listed in the previous step, there should be one entity that attracts other entities towards it as it has the most exchange flows to share. The anchor entity is the largest giver in the EIP. It is important to identify the anchor entity to ensure that industrial symbiosis between entities continues.

DETERMINE INDUSTRIAL SYMBIOSIS

Exchange flows between entities should be listed and determined. This visualizes all possible industrial symbiosis in the EIP. For simplicity, we propose the EIP matrix that lists all industrial symbiosis in the EIP. The EIP matrix summarizes all exchange flows and gives detailed information about the network.

DEFINE EXCHANGE-FLOW TYPES

Through a graphical presentation, Al-Quradaghi et al. (2020) highlight the different exchange types on the basis of the literature. The types of exchange flow can be defined in this step. In this chapter, we considered external-exchange Types 3–5 as the main flow types to build an EIP. For industrial symbiosis to happen, there needs to be external and not internal exchange flow. We define "internal exchange" as any industrial symbiosis that exists within the entity (e.g., old equipment from one department in the entity can be used in another department). On the other hand, "external exchange" is any industrial symbiosis that appears beyond the boundary of the entity (e.g., old equipment sent to recycling company), including co-located, non-co-located, and regional firms. This requirement is very important, as it identifies the distance between entities and the modes of transportation of the exchange flows (trucks, pipelines, etc.).

SUSTAINABILITY FRAMEWORK

ELV OVERVIEW

The European Directive of ELVs 2000/53/EC defines end-of-life vehicles as "vehicles that have become waste," and waste defined as "any substance or object which the holder discards, or intends to discard, or is requires to discard" [35]. According to the Official Journal of the European Communities, end-of-life vehicles account for up to 10% of the total amount of waste generated annually in the European Union [35]. The directive requires car manufacturers to ensure that a minimum of 95% by weight per vehicle is reusable and/or recyclable, including a minimum of 85% of material recoverability (recyclability) or reuse [36].

The waste stream generated from end-of-life vehicles can be controlled/regulated through the vehicle-design phase when product development occurs. "Design for X" is one of several methods to aid the designer in this phase, where "X" refers to the life-cycle phase being evaluated [37–40]. In other words, "X" represents the aim of the design: recycling, quality, sustainability, cost, and so on. It requires the design to meet the specific defined goal (X).

One solution could be introducing design for sustainability (DfS) in the design phase that requires maximizing resource efficiency while minimizing environmental impact [29, 40, 41]. A more specific definition for DfS is given by Vezzoli et al. (2018): "A design practice, education, and research that, in one way or another, contributes to sustainable development" [42].

Using recycled scraps benefits the environment threefold by: (i) saving raw materials, (ii) saving energy, and (iii) reducing GHG emissions [29, 41, 43]. Taking steel as an example to demonstrate the use of recycling, every ton of new steel made from scrap steel saves 2500 lb of iron ore, 1400 lb of coal, and 120 lb of limestone (Fold i). The use of recycled scrap steel reduces energy use by 75% (Fold ii). The estimated GHG reduction for recycling steel in every recycled vehicle is 2205 lb of GHGs (Fold iii; [44]). In general, energy used in recycling scrap materials is less than the energy used in manufacturing raw materials [40, 45]. Energy saving and GHG reduction [44] for some metals are illustrated in the case study (see Al-Quradaghi et al., 2020).

However, the dilemma is to figure out if recycling ELVs is economically feasible. Villalba et al. (2002) proposed a recyclability index to measure "the ability of a material to regain its valued properties through recycling process" [45]. The recyclability index calculates the profit-to-loss margin for recycling; hence, it determines whether it is economically feasible to recover the material [29]. With a positive margin, recycling is a good choice; a negative margin indicates some concerns making the material not worth recycling.

The end-of-life-vehicle recycling system aims to isolate hazardous content, and recover usable parts and recycle others [43]. There is a tremendous number of studies in the literature about recycling systems and managing ELVs that were comprehensively studied by [46, 47].

Vehicles mainly go through different stages. Maudet et al. (2012) highlighted two main systems for treating ELVs, dismantling components and vehicle shredding [48]. In more detail, Edwards et al. (2006) described three main stages in the recycling process: (1) depollution, (2) dismantling, and (3) shredding [49]. The case study report illustrates the overall stages for treating ELVs, including depollution, dismantling, and shredding (see Al-Quradaghi et al., 2020). The elements are summarized below:

- Depollution: Drain all fluids (gas, oil, coolant, etc.) and remove battery.
- Dismantling: Remove engine, tires, windshield, and steering wheel.
- Shredding: Press the hulk and send to shredding machine.

In the first stage (depollution), all fluids are drained and the battery is removed. The second stage (dismantling) removes the engine, tires, wires, cables, windows, bumpers, and other parts that are useable. In the third stage (shredding), the vehicle's hulk is pressed using a hammermill and then sent to the shredding machine. In the process, ferrous metal is separated using magnetic separation, and an eddy current is used to separate the nonferrous metal. The remainder of that process (plastics, rubber, fabrics, and dirt) is called automobile shredder residue (ASR) or fluff [40, 43, 49–51]. According to Curlee et al. (1994), ASR generated from recycling ELVs accounts for about 25% (by weight) of the shredded material [40]. The highest percentage of components in the ASR is fibers, which accounts for 42% by weight, followed by plastics, with 19.3% by weight. There are several approaches to separate the components or use the ASR for different purposes [52, 53].

ELV management is a crucial issue to deal with for governments, vehicle producers, and treatment facilities. It has received increased attention due to its implications, both economic and environmental. The problem has both tactical- and strategic-level decision-making components. D'Adamo et al. (2020) prepared regression models to predict the amount of ELVs generated yearly as a function of GDP and population. They concluded that, given the great amount of ELVs generated, adopting a practical procedure for constructing efficient procedures to connect and induce collaborations between the actors involved in ELV will help greatly in enhancing sustainability and creating economic opportunities [54]. Karagoz et al. provided a comprehensive review of 232 peer-reviewed articles published in the period 2000–2019 that was aimed at identifying the gaps in the ELV management literature [47]. They concluded that only few researchers suggested solutions that closed the waste management loop by recycling and suggested that such approaches should be devised for the solution of real-life ELV management problems to generate reasonable solutions for them. Finally, in a recent article that provided a bibliometric literature review and assessed the efficiency of ELV management, De Almeida and Borsato (2019) concluded that the literature reveals a series of strategies that are confusing [55]. The paper outlined several ELV management strategies and the different processes involved. The paper also concluded that waste management research focusing on the holistic nature of the ELV problem and which considers nodes of different efficiencies is still lacking. Furthermore, the paper suggested future research management strategies that focus on sustainability and the triple bottom line. Therefore, the proposed strategy of the previous section is clearly a step forward towards bridging this gap and its use is illustrated on the case of the ELV management problem in the next sections.

The proposed methodology must also be tailored to the type of vehicles that are in existence in a certain country and also to the specifics of that country. For instance, Che et al. (2011) discuss the specifics of ELV problem in Japan, China, and Korea, and propose different scenarios [56]. For example, labor cost is high in Japan and the design of an EIP (applied to the ELV management problem) must take this into account. This can be done by focusing on automation and taking the economics of the problem into consideration. For example, in the case of several possible alternatives available in one of the suggested nodes by the proposed methodology, the different scenarios must be compared based on a composite objective that considers both the NPV and the sustainability component. In this way, the most desirable alternative with respect to this composite objective is selected. The specific nature of a given country can also be in terms of enforced recycling laws. Step 2 of the methodology that focuses on the identification of all entities with industrial symbiosis must therefore be altered to consider only entities that conform to the law of the country where the recycling unit is to be implemented.

New generation vehicles (NGVs), like hybrid, plug-in hybrid, and electric vehicles, are emerging into the market with increasing rates due to advances in battery technology, material design, and computerized technology. These vehicles have different components compared to the traditional fossil fuel vehicles. They are equipped, for example, with highly efficient nickel-hydrogen or lithium-ion batteries. When the proposed methodology is applied to NGVs, in Step 2, which is concerned with the identification of all entities with industrial symbiosis, should take into account the balance between reuse and recycling. For NGVs, there is an emerging trend for the effective utilization of waste batteries [57]. Furthermore, because of the inclusion of these highly efficient batteries, the steel content of the vehicles is much lower than that of the traditional fossil fuel vehicles while plastic content is more. For these reasons, nontraditional processing and recycle nodes should be considered for the case of NGVs. Yu et al. (2017) provides a comprehensive analysis of the different recycle and reuse approaches of waste batteries from NGVs [57].

ECO-PARK FOR END-OF-LIFE VEHICLES

The proposed framework was applied to solve end-of-life vehicles (EIP-4-ELVs). The steps of the framework tailored to this case are illustrated in the case study report (see Al-Quradaghi et al., 2020). The motivations for the case of EIP-4-ELVs are environmental and economic. End-of-life vehicles are harmful to the environment, so solving this problem is the main motivation. Furthermore, recycling old cars generates profit for many industries. The next step is to identify which industries form the EIP-4-ELVs. The entities are suggested to be power plants, dismantling facilities, waste-to-energy plants, wastewater-treatment plants, glass industries, tire recycling, aluminum, plastic, and steel companies, and battery-recycling or -refurbishing companies. The anchor entity that generates the most waste/by-products in the case of EIP-4-ELVs is the dismantling facility. This facility is the core of EIP-4-ELVs as it sends out scrap materials of different types to the corresponding industries. Next is to determine all possible exchange flows in the EIP-4-ELVs using the EIP matrix. Last is to define the exchange-flow type for transportation purposes—External Type 5 in this case. The feedback loop between the last two steps allows for any future change or modification in the exchange flow. The EIP matrix is provided for this case in a tabulated format (see Al-Quradaghi et al., 2020). Cells with (-) in the table indicate possible industrial symbiosis in the future as the EIP evolves.

The anchor entity, as mentioned earlier, is the dismantling facility (DF), which is the main entity for sending by-products/waste to other industries. The waste and multiple by-products created from the DF are considered resources for the other industries. The proposed industrial symbiosis given by the EIP matrix is shown in a connected network presented in the study report (see Al-Quradaghi et al., 2020).

Nodes represent industries in the EIP-4-ELVs, and arcs show the exchange flows. The network illustrates how entities in the proposed EIP-4-ELVs can utilize waste from each other. In the next section, the case-study authors ran a simulation model to the developed network and present the outcomes. As the aim of this chapter was to solve ELVs, the focus was to study the material flows in the proposed system boundary.

EIP-4-ELV SIMULATION

MODEL ASSUMPTIONS

The vehicle-dismantling facility (DF) is the source for all types of vehicles. In this study, we based our calculations on data given by the United Nations Environment Program (UNEP) in the 2013 report titled *Metal Recycling, Opportunities, Limits, Infrastructure* [41], and from the original source, the European Commission Joint Research Center report, "Environmental Improvement of Passenger Cars" (See Al-Quradaghi et al, 2020). We considered two types of passenger vehicles: petrol and diesel. Al-Quradaghi et al. (2020) present the composition of an average passenger car from each type, showing the average curb weights for each type. As defined by the U.S. Department of Transportation, curb weight is "the actual weight of the vehicle with a full tank of fuel and other fluids needed for travel, but no occupants or cargo" [58].

In the available data, this did not add up to the total weight because of the lack of detailed information [58]. For the purpose of this chapter, we did the following: (1) added Paint and Textile categories to the other category; (2) calculated the percentage of material; and (3) calculated the average material composition from the two types, and used it instead of weight. The average material composition was tabulated in the study report (see Al-Quradaghi et al., 2020).

In implementing the model, the following assumptions were made: (1) the dismantling facility operates 10 hours per day (7:00 to 17:00); (2) three Powerhand vehicle-recycling-system (VRS) machines are used to dismantle the vehicles; and (3) average processing time for each vehicle is known.

MODEL RESULTS

The connected network of exchange flows is simulated via the SIMIO software package [59]. The produced results spanning one month (28 working days) are summarized below:

- Iron and Steel: 1,548
- Aluminum: 127
- Glass: 73
- Plastic: 207
- Tires: 56
- Battery: 25
- Fluids: 91
- Other: 157

As can be seen in the table, the dismantling facility processed 1820 vehicles, and the material outcome from the network could be used for each corresponding industry as a resource. This is only an example of how one month of recycling ELVs produces a different quantity of materials that can be reused. If done for the long term, the EIP-4-ELV network serves in the reduction of raw material extractions and GHG emissions.

CONCLUSIONS

Eco-industrial parks (EIPs) promote the shift from the traditional linear to the circular model, where by-products and waste can be reused. The EIP literature covers multidisciplinary areas, including optimization, life-cycle assessments, policy implementation, social networks, and typology. The International Framework for EIP provided by the UNIDO, GIZ, and WBG report serves as a guide for decision-makers on important components to achieve maximal benefits economically, environmentally, and socially. However, as indicated in the report, the International Framework for EIP provides only strategic details for the EIP requirements, and does not translate them into an existing EIP.

On the other hand, the frameworks for EIPs that are provided in the literature are very comprehensive. Hence, there is increasing need for a straightforward framework. The proposed framework illustrates the foundations of designing eco-industrial parks. It shows general step-by-step actions to be discussed at the very early stages of designing EIPs. The framework answers fundamental questions that decision-makers need to consider for developing and designing EIPs. In a simple, clear, and step-by-step strategy, the framework lays out necessary actors in the process of designing EIPs. The framework requires primary data that can easily be collected from each entity that forms EIPs.

The proposed framework bridges the gap in the literature and provides a generalized framework for the design of eco-industrial parks. The framework was employed to solve the end-of-life-vehicle problem (EIP-4-ELVs). As a result of applying the framework in EIP-4-ELVs, a connected network of exchanges was built. The outcomes represent the amount of different types of materials. If applied to solve the end-of-life-vehicle problem, the framework can create a connected network that produces different types of materials. By using EIP-4-ELVs, the network prevents using more natural sources, and depends on some percentage of the by-product exchange from other industries.

REFERENCES

1. Frosch, R. A., and N. E. Gallopoulos 1994, "Strategies for Manufacturing," *Scientific American*, Vol. 261, pp. 144–153.
2. Frosch, R. A. 1994, "Industrial Ecology: Minimizing the Impact of Industrial Waste," *Physics Today*, Vol. 47, pp. 63–68.
3. Gu, C., S. Leveneur, and L. Estel Yassine, A. 2013, "Modeling and Optimization of Material/Energy Flow Exchanges in an Eco-Industrial Park," *Energy Procedia*, Vol. 36, 243–252.
4. Chertow, M. R. 2007, "Uncovering' Industrial Symbiosis," *Journal of Industrial Ecology*, Vol. 11, p. 20.

5. Ehrenfeld, J., and N. Gertler 1997, "Industrial Ecology in Practice; The Evolution of Interdependence at Kalundborg," *Journal of Industrial Ecology*, Vol. 1, pp. 67–79.

6. Piaszczyk, C. 2011, "Model Based Systems Engineering with Department of Defense Architectural Framework," *Systems Engineering*, Vol. 14, pp. 305–326.

7. Behera, S. K., J.-H. Kim, S.-Y. Lee, S. Suh, and H.-S. Park 2012, "Evolution of 'Designed' Industrial Symbiosis Networks in the Ulsan Eco-Industrial Park: 'Research and Development into Business' as the Enabling Framework," *Journal of Cleaner Production*, Vol. 29–30, pp. 103–112.

8. Tessitore, S., T. Daddi, and F. Iraldo 2015, "Eco-Industrial Parks Development and Integrated Management Challenges: Findings from Italy," *Sustainability*, Vol. 7, pp. 10036–10051.

9. Mat, N., J. Cerceau, L. Shi, H. S. Park, G. Junqua, and M. Lopez-Ferber 2016, "Socio-Ecological Transitions Toward Low-Carbon Port Cities: Trends, Changes and Adaptation Processes in Asia and Europe," *Journal of Cleaner Production*, Vol. 114, pp. 362–375.

10. Aid, G., M. Eklund, S. Anderberg, and L. Baas 2017, "Expanding Roles for the Swedish Waste Management Sector in Inter-Organizational Resource Management," *Resources, Conservation & Recycling*, Vol. 124, pp.85–97.

11. Susur, E., A. Hidalgo, and D. Chiaroni 2019, "A Strategic Niche Management Perspective on Transitions to Eco-Industrial Park Development: A Systematic Review of Case Studies," *Resources, Conservation & Recycling*, Vol. 140, pp. 338–359.

12. Boix, M., L. Montastruc, C. Azzaro-Pantel, and S. Domenech 2015, "Optimization Methods Applied to the Design of Eco-Industrial Parks: A Literature Review," *Journal of Cleaner Production*, Vol. 87, pp. 303–317.

13. Zhang, Y., S. Duan, J. Li, S. Shao, W. Wang, and S. Zhang 2017, "Life Cycle Assessment of Industrial Symbiosis in Songmudao Chemical Industrial Park, Dalian, China," *Journal of Cleaner Production*, Vol. 158, pp. 192–199.

14. Jiao, W., F. Boons 2014, "Toward a Research Agenda for Policy Intervention and Facilitation to Enhance Industrial Symbiosis Based on a Comprehensive Literature Review," *Journal of Cleaner Production*, Vol. 67, pp. 14–25.

15. Song, X., Y. Geng, H. Dong, and W. Chen 2018, "Social Network Analysis on Industrial Symbiosis: A Case of Gujiao Eco-Industrial Park," *Journal of Cleaner Production*, Vol. 193, pp. 414–423.

16. UNIDO; GIZ; WBG 2017, *An International Framework for Eco-Industrial Parks*, UNIDO, Danver, MA, USA.

17. Haskins, C. 2007, "A Systems Engineering Framework for Eco-Industrial Park Formation," *Systems Engineering*, Vol. 10, pp. 83–97.

18. Sopha, B. M., A. M. Fet, M. M. Keitsch, and C. Haskins 2009, "Using Systems Engineering to Create a Framework for Evaluating Industrial Symbiosis Options," *Systems Engineering*, Vol. 13, pp. 149–160.

19. Boons, F., W. Spekkink, and Y. Mouzakitis 2011, "The Dynamics of Industrial Symbiosis: A Proposal for a Conceptual Framework Based Upon a Comprehensive Literature Review," *Journal of Cleaner Production*, Vol. 19, pp. 905–911.

20. Dumoulin, F., T. Wassenaar, A. Avadi, and J. Paillat 2016, "A Framework for Accurately Informing Facilitated Regional Industrial Symbioses on Environmental Consequences," *Journal of Industrial Ecology*, Vol. 21, pp. 1049–1067.

21. Konstantinova, Y., E. Johannes, and B. Vejrum 2019, "Dare to Make Investments in Industrial Symbiosis? A Conceptual Framework and Research Agenda for Developing Trust," *Journal of Cleaner Production* Vol. 223, pp. 989–997.

22. Andiappan, V., R. R. Tan, and D. K. S. Ng 2016, "An optimization-Based Negotiation Framework for Energy Systems in an Eco-Industrial Park," *Journal of Cleaner Production*, Vol. 129, pp. 496–507.

23. Yedla, S., and H. Park 2017, "Eco-Industrial Networking for Sustainable Development: Review of Issues and Development Strategies," *Clean Technologies and Environmental Policy*, Vol. 19, pp. 391–402.

24. Romero, E., and M. C. Ruiz 2013, "Framework for Applying a Complex Adaptive System Approach to Model the Operation of Eco-Industrial Parks," *Journal of Industrial Ecology*, Vol. 17, pp. 731–741.

25. Tao, Y., S. Evans, Z. Wen, and M. Ma 2019, "The Influence of Policy on Industrial Symbiosis from the Firm's Perspective: A Framework," *Journal of Cleaner Production*, Vol. 213, pp. 1172–1187.

26. Liu, C., and R. Côté 2017, "A Framework for Integrating Ecosystem Services into China's Circular Economy: The Case of Eco-Industrial Parks," *Sustainability*, Vol. 9, p. 1510.

27. Gopinath, A., A. Bahurudeen, S. Appari, and P. Nanthagopalan 2018, "A Circular Framework for the Valorisation of Sugar Industry Wastes: Review on the Industrial Symbiosis Between Sugar, Construction and Energy Industries," *Journal of Cleaner Production*, Vol. 203, pp. 89–108.

28. Kuznetsova, E., E. Zio, and R. Farel 2016, "A Methodological Framework for Eco-Industrial Park Design and Optimization," *Journal of Cleaner Production*, Vol. 126, pp. 308–324.

29. Villalba, G., M. Segarra, J. M. Chimenos, and F. Espiell 2004, "Using the Recyclability Index of Materials as a Tool for Design for Disassembly," *Ecological Economics*, Vol. 50, pp. 195–200.

30. Chertow, M. R. 2000, "Industrial Symbiosis: Literature and Taxonomy," *Annual Review of Environment and Resources*, Vol. 25, pp. 313–337.

31. Eilering, J. A. M., and Vermeulen, W. J. V. 2004, "Eco-Industrial Parks: Toward Industrial Symbiosis and Utility Sharing in Practice." *Progress in Industrial Ecology, An International Journal*, Vol. 1, p. 245.

32. Felicio, M., D. Amaral, K. Esposto, and X. G. Durany 2016, "Industrial Symbiosis Indicators to Manage Eco-Industrial Parks as Dynamic Systems," *Journal of Cleaner Production*, Vol. 118, pp. 54–64.

33. Heeres, R. R., W. J. V. Vermeulen, and F. B. de Walle 2004, "Eco-Industrial Park Initiatives in the USA and the Netherlands: First Lessons," *Journal of Cleaner Production*, Vol. 12, pp. 985–995.

34. Tian, J., W. Liu, B. Lai, X. Li, and L. Chen 2014, "Study of the Performance of Eco-Industrial Park Development in China," *Journal of Cleaner Production*, Vol. 64, pp. 486–494.

35. European Parliament and Council 2000, "Directive 2000/53/EC on End-of-Life Vehicles," *Official Journal of the European Communities European Parliament and Council: Brussels, Belgium*, Vol. L269, pp. 34–42.

36. Garcia, J., D. Millet, and P. Tonnelier 2015, "A Tool to Evaluate the Impacts of an Innovation on a Product's Recyclability Rate by Adopting a Modular Approach: Automotive Sector Application," *International Journal of Vehicle Design*, Vol. 67, pp. 178–204.

37. Watson, B., and D. Radcliffe 2010, "Structuring Design for X Tool Use for Improved Utilization," *Journal of Engineering, Design*, Vol. 9, pp. 211–223.

38. Meerkamm, H. 2007, "Design for X—A Core Area of Design Methodology," *Journal of Engineering, Design*, Vol. 5, pp. 165–181.

39. Tonnelier, P., D. Millet, S. Richir, and M. Lecoq 2007, "Is It Possible to Evaluate the Recovery Potential Earlier in the Design Process? Proposal of a Qualitative Evaluation Tool," *Journal of Engineering, Design*, Vol. 16, pp. 297–309.

40. Curlee, T. R., S. Das, C. G. Rizy, and S. M. Schexnayder 1994, *Recent Trends in Automobile Recycling: An Energy and Economic Assessment*, Oak Ridge National Laboratory, Oak Ridge, TN, USA.
41. Reuter, M. A., C. Hudson, A. van Schaik, K. Heiskanen, C. Meskers, and C. Hagelüken 2013, *Metal Recycling: Opportunities, Limits, Infrastructure, a Report of the Working Group on the Global Metal Flows to the International Resource Panel*, UNEP, Paris, France.
42. Vezzoli, C., F. Ceschin, L. Osanjo, M. K. M'Rithaa, R. Moalosi, V. Nakazibwe, and J. C. Diehl 2018, *Designing Sustainable Energy for All. Sustainable Product-Service System Design Applied to Distributed Renewable Energy*, Springer, Cham, Switzerland, p. 230.
43. Andersson, M. 2016, *Innovating Recycling of End-of-Life Cars*, Chalmers University of Technology, Göteborg, Sweden.
44. ARA 2012, *Automotive Recycling Industry: Environmentally Friendly, Market Driven, and Sustainable*, ARA, Manassas, VA, USA.
45. Graedel, T. E., J. Allwood, J.-P. Birat, B. K. Reck, S. F. Sibley, G. Sonnemann, M. Buchert, and C. Hagelüken 2011, *Recycling Rates of Metals—A Status Report, a Report of the Working Group on the Global Metal Flows to the International Resource Panel*, UNEP, Paris, France.
46. Simic, V. 2013, "End-of-Life Vehicle Recycling—A Review of the State-of-the-Art," *Recikliranje vozila na Kraj. životnog ciklusa—Pregl. Najsuvremnijih Znan. Rad.*, Vol. 20, pp. 371–380.
47. Karagoz, S., N. Aydin, and V. Simic 2019, "End-of-Life Vehicle Management: A Comprehensive Review," *Journal of Material Cycles and Waste Management*, Vol. 22, pp. 416–442.
48. Maudet, C., G. Yannou-Le Bris, and D. Froelich 2012, "Integrating Plastic Recycling Industries into the Automotive Supply Chain," *HAL*, Vol. 13, pp. 71–89.
49. Edwards, C., T. Bhamra, and S. Rahimifard 2006, "A DesignFrameworkforEnd-of-LifeVehicleRecovery," *Proceedings of the 13th CIRP International Conference on Life Cycle Engineering*, Leuven, Belgium, 31 May–2 June; pp. 365–370.
50. Choi, J. K., J. A. Stuart, and K. Ramani 2005, "Modeling of Automotive Recycling Planning in the United States," *International Journal of Automotive Technology*, Vol. 6, pp. 413–419.
51. Paul, R. 2009, "End-of-Life Management of Waste Automotive Materials and Efforts to Improve Sustainability in North America," *WIT Transactions on Ecology and the Environment*, Vol. 120, pp. 853–861.
52. Baker, B. A., N. J. Brookside, K. L. Woodruff, P. Morrisville, J. F. Naporano, and N. J. E. Fells 1995, Automobile Shredder Residue (ASR) Separation and Recycling System. U.S. Patent WO1995026826A1, 12 October.
53. van Schaik, A., and M. A. Reuter 2004, "The Optimization of End-of-Life Vehicle Recycling in the European Union," *JOM*, Vol. 56, pp. 39–43.
54. D'Adamo, I., M. Gastaldi, and P. Rosa 2020, "Recycling of End-of-Life Vehicles: Assessing Trends and Performances in Europe," *Technological Forecasting and Social Change*, Vol. 152, p. 119887.
55. de Almeida, S. T., and M. Borsato 2019, "Assessing the Efficiency of End of Life Technology in Waste Treatment—A Bibliometric Literature Review," *Resources, Conservation & Recycling*, Vol. 140, pp. 189–208.
56. Che, J., J. Yu, and R. S. Kevin 2011, "End-of-Life Vehicle Recycling and International Cooperation between Japan, China and Korea: Present and Future Scenario Analysis," *Journal of Environmental Sciences*, Vol. 23, pp. S162–S166.

57. Yu, J., S. Wang, K. Toshiki, K. R. B. Serrona, G. Fan, and B. Erdenedalai 2017, Latest Trends and New Challenges in End-of-life Vehicle Recycling. In *Environmental Impacts of Road Vehicles: Past, Present and Future*, J. Yu, editor. The Royal Society of Chemistry, London, UK, Volume 44, pp. 174–213.

58. Kahane, C. J. 2003, *Vehicle Weight, Fatality Risk and Crash Compatibility of Model Year 1991–99 Passenger Cars and Light Trucks*, National Highway Traffic Safety Administration, Springfield, VA, USA.

59. Joines, J. A., and S. Roberts 2015, *Simulation Modeling with SIMIO: A Workbook*, 4th edition, SIMIO LLC, Sewickley, PA, USA.

60. Al-Quradaghi, Shimaa, Quipeng P. Zheng, and Ali Eklamel 2020, "Generalized Framework for the Design of Eco-Industrial Parks: Case Study of End-of-Life Vehicles," *Sustainability*, Vol. 12, p. 6612. Doi: 10.3390/su12166612.

9 Climate Change Uncertainty Modeling

REPRINTED WITH OPEN ACCESS PERMISSION FROM

Weiss, Scott C., Justin D. Delorit, and Christopher M. Chini (2022), "Energy Forecasting to Benchmark for Federal Net-Zero Objectives Under Climate Uncertainty," *Environmental Research: Infrastructure and Sustainability*, Vol. 2, No. 4 (Resource Consumption and Sustainability in the Built Environment), pp. 1–14. DOI: 10.1088/2634-4505/ac9712

INTRODUCTION

Climate variability creates energy demand uncertainty and complicates long-term asset management and budget planning. Without understanding future energy demand trends related to intensification of climate, changes to energy consumption could result in budget escalation. Energy demand trends can inform campus infrastructure repair and modernization plans, effective energy use reduction policies, or renewable energy resource implementation decisions, all of which are targeted at mitigating energy cost escalation and variability. To make these long-term management decisions, energy managers require unbiased and accurate energy use forecasts. This research uses a statistical, model-based forecast framework, calibrated retrospectively with open-source climate data, and run in a forecast mode with CMIP5 projections of temperature for RCPs 4.5 and 8.5 to predict total daily energy consumption and costs for a campus-sized community (population: 30 000) through the end of the century. The case study of Wright Patterson Air Force Base is contextualized within the existing executive orders directing net-zero emissions and carbon-free electricity benchmarks for the federal government. The model suggests that median annual campus electric consumption, based on temperature rise alone, could increase by 4.8% with RCP4.5 and 19.3% with RCP8.5 by the end of the century, with a current carbon footprint of 547 million kg CO_2. Monthly forecasts indicate that summer month energy consumption could significantly increase within the first decade (2020–2030), and nearly all months will experience significant increases by the end of the century. Therefore, careful planning is needed to meet net-zero emissions targets with significant increases in electricity demands under current conditions. Policies and projects to reduce the carbon footprint of federal agencies need to incorporate forecasting models to understand changes in demand to appropriately size electric infrastructure.

Climate variability is an exogenous, stochastic factor that causes energy demand uncertainty and complicates energy consumption modeling. Limited understanding of future energy demand trends can leave energy managers ill-prepared to make long-term decisions, such as whether to advocate for infrastructure modernization or expansion, make facilities more energy-efficient, or consider renewable energy resources. Energy managers require forecast models to project future energy demand and inform these decisions and their budgets. Managing energy at the campus level is particularly difficult because resource allocation must be prioritized among many facilities (Kim et al. 2019). Compounding this energy management challenge is a growing need and demand for transition to renewables and carbon-free energy. In particular, the Department of Defense has been directed by recent executive order (EO) 14057 to achieve 100 percent carbon-free electricity on a net annual basis by 2030 and net-zero emissions by 2050. This places a burden on energy managers to transition energy production while dealing with an increasing demand. In this study we analyze the existing trends of electric consumption at Wright Patterson Air Force Base (WPAFB), build and validate a forecast model for future consumption prediction, and discuss these implications in the context of decarbonization.

Gradual changes to climate and its forecasted impact on energy consumption is a key area of research with important implications for energy management and renewable transitions (De Felice et al. 2015; Dirks et al. 2015; Emodi et al. 2018; Moral-Carcedo and Pérez-García 2019; Mukherjee et al. 2019; Son and Kim 2017; Zhou et al. 2014). For example, Emodi et al. (2018) forecast electrical energy consumption in Australia using climate projections and find that electrical energy consumption may decrease in the first few decades due to reduced heating demand before increasing and surpassing current energy consumption levels by the end of the 21st century as cooling demands rise. Zhou et al. (2014) forecast the impact of changing climate across the 21st century for United States heating and cooling, finding that heating energy demand will gradually drop for all states, but cooling energy demand will ultimately increase after decreasing slightly between 2005 and 2020. Importantly, this shift in demand will further increase summer peak demands of electricity, where fossil fuel power plants are already stressed due to drought or heatwaves, reducing cooling efficiency (McCall and Macknick 2016, Cook et al. 2015, Chowdhury et al. 2021).

Energy prediction and forecast analyses exist at the facility, regional, and multinational scales (Al-bayaty et al. 2019; Amato et al. 2005; De Rosa et al. 2014; Mukherjee et al. 2019; Mukherjee and Nateghi 2017; Xie and Hong 2017). For example, De Rosa et al. (2014), at the facility level, addresses how energy savings policy can impact energy consumption, while Amato et al. (2005) provide a regional-level analysis that focuses on how changes to energy infrastructure impact energy consumption spatially. However, at an organization or campus-level, few studies exist, though aggregation or disaggregation methods are most commonly used to achieve results at this level of analysis. Dirks et al. (2015) acquire facility energy modeling software that houses thousands of facility types, and energy information on an entire geographical region of the US to model energy consumption, and inform energy mitigation and savings techniques. The difficulty of campus-level analyses lies in capturing the different use regimes between facilities along with accessing large quantities of facility-level data.

To meet the objectives of recent EOs, including EO 14057, the DoD must consider how projected changes in climate may impact energy demand and use patterns.

Accurate projections of future energy costs are essential for informing short- and long-term decisions and budgets and for meeting renewable energy goals. For instance, in the short term, accurate year-ahead forecasts can prevent underbudgeting, which drives the need to borrow from other facility sustainment funds. In the long term, they inform long-range organizational operating budgets, facilitate energy management policy, and motivate investments in energy infrastructure. In this study, we develop and validate a forecasting model based on historical energy demand and climate variables to predict future electric demand based on multiple climate change scenarios. We utilize data from WPAFB as a case study to showcase the increasing energy demands of the Department of Defense. Additionally, we discuss the current trends in renewable energy in southeastern Ohio, where Wright-Patterson Air Force Base is located.

BACKGROUND

Statistical, climate-driven prediction models have been used across many fields to gain insight into past and future impacts on operations and to inform policy. Models applied to the management of built and natural systems are applied broadly and produce results with varying degrees of deterministic and probabilistic skill (Delorit et al. 2017; Graafland et al. 2020; Theusme et al. 2021; Zeng et al. 2020). Models have been developed for the energy sector at a wide range of temporal, spatial, and organizational scales, and are most commonly calibrated to evaluate energy consumption with mention of climate impacts (Amato et al. 2005; Apadula et al. 2012; De Felice et al. 2015; De Rosa et al. 2014; Dirks et al. 2015; Emodi et al. 2018; Mansur et al. 2008; Moral-Carcedo and Pérez-García 2019; Mukherjee et al. 2019; Mukherjee and Nateghi 2017; Son and Kim 2017; Zhou et al. 2014).

Climate-driven, empirical statistical prediction models are developed using a variety of climate inputs, spatial scales, and regression techniques. Many climate variables have been shown to provide value in energy consumption prediction models. In cases of limited access to data, variables are selected based on intuition or expertise in a specific area (Son and Kim 2017). Alternatively, the ability to perform exhaustive analyses may be limited considering that energy managers are typically not modelers or climate scientists. However, existing literature has highlighted specific key climate variables that may help researchers limit their search space, as summarized in Table 9.1. This study builds on these previous works and investigates a multitude of climate factors for their explanatory power in predicting energy use.

Similar to facility-level models, state- or region-level models can be exported to other states and regions to test the model's stability under a diverse array of climate conditions (Mukherjee and Nateghi 2017; Mukherjee et al. 2019). Additionally, state-level data can be disaggregated to better understand business sector energy consumption and to capture the spatial heterogeneity of building use within each state (Zhou et al. 2014). For national or multinational scale, the difficulty lies in data collection. Wenz et al. (2017) gather electrical energy consumption data from across Europe to develop their wide-ranging study and provided a better understanding of Europe's predicted peak energy consumption under climate change.

Finally, Chandramowli and Felder (2014) present a review of energy consumption prediction methods that found multiple linear regression to be one of the

TABLE 9.1
Previous Research Investigated Energy Demand as a Function of Several Climatological Variables

Climate variable	References	Notes
Temperature	Al-bayaty et al. (2019), Amato et al. (2005), Apadula et al. (2012), De Felice et al. (2015), Dirks et al. (2015), Ismail and Abdullah (2016), Mukherjee et al. (2019), Mukherjee and Nateghi (2017), Psiloglou et al. (2009), Xie and Hong (2017), Zhou et al. (2014)	Relationship depends on prevalence of electric space conditioning; cooler climates see more linear relationship (negative); warmer climates see a non linear relationship with a higher demand at high temperatures; cooling/heating degree days often used instead
Relative humidity	Al-bayaty et al. (2019), Apadula et al. (2012), Dirks et al. (2015)	Slightly positive, especially in conjunction with high temperatures; also modeled with heat index
Cloud cover	Al-bayaty et al. (2019), Apadula et al. (2012)	Minimal significance, slightly positive due to increased lighting
Precipitation	Fan et al. (2015), Mansur et al. (2008), Mukherjee et al. (2019), Mukherjee and Nateghi (2017)	Minimal statistical correlation with energy demand
Wind speed	Al-bayaty et al. (2019), Mukherjee et al. (2019), Mukherjee and Nateghi (2017), Xie and Hong (2017)	Partial correlation with more sensitivity in the residential sector
Irradiation	Al-bayaty et al. (2019), De Rosa et al. (2014)	Significant in warmer climates with cooling demands

prominent techniques. Other technique types include fuzzy regression (Chukhrova and Johannssen 2019), Bayesian additive regression trees (Mukherjee et al. 2019; Mukherjee and Nateghi 2017), support vector regression (De Felice et al. 2015; Son and Kim 2017), and artificial neural networks (Al-bayaty et al. 2019; Ismail and Abdullah 2016). However, no studies have leveraged principal component analysis (PCA) with regression, much less with cross-validated multiple linear regression, to account for multicollinearity and bias present in climate and other predictors.

METHODS

DATA

For this study, electric consumption data were provided by WPAFB, located near Dayton, OH, across four consecutive years at the hourly scale, in kilowatt-hours (1 Oct. 2015–30 Sep. 2019). The scale of these data most closely resemble that of a city, manufacturing complex, or medical or university campus. WPAFB employs over 30,000 people and includes various operation types such as, industrial, commercial, community support, and residential. In all, the data include energy demand from approximately 26,500 facilities. Dayton, Ohio has a temperate climate with moderate rainfall throughout the year, warm to hot summers, and cool to cold winters. While the authors are not permitted to supply billed consumption, or rates, they may be requested via open access request from the installation.

A majority of the climate data used in this prediction framework was retrieved from the National Oceanic and Atmospheric Administration's (NOAA) Local Climatological Data database to include dry bulb temperature, wet bulb temperature, dew point temperature, relative humidity, station pressure, sea pressure, wind speed, precipitation, and cloud fraction. Solar irradiation, cloud opacity, and precipitable water variables were obtained from the commercial solar forecasting company Solcast (Solcast 2019).

Open-source climate projections were obtained through the Lawrence Livermore National Laboratory website (Maurer et al. 2007). All available models and ensembles for the CMIP5 bias-corrected daily climate projections (BCCAv2-CMIP5-Climate-daily) for maximum and minimum temperature were selected for both RCP 4.5 and RCP8.5 (Reclamation 2013). The two RCPs are used to demonstrate two potential ranges of future energy consumption values. The projection set for RCP4.5 consists of 19 models, with a total of 42 projection ensembles, and RCP8.5 consists of 20 models, with a total of 41 projection ensembles. The median value for all ensembles and each RCP was used to consolidate the ensembles into a single set of deterministic predictions. This approach is consistent with many studies where capturing the general responses to climate change is desired. Figure 9.1 shows the

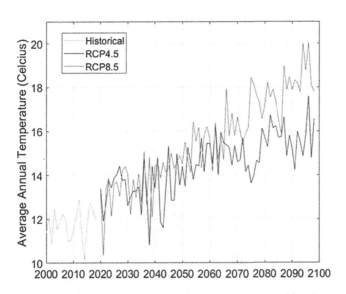

FIGURE 9.1 Time series of the historical and forecasted average yearly temperatures. Generally, temperature will gradually increase throughout the century. Between years 2010 and 2040, there is a period of less substantial temperature increase, where a few years may experience average yearly temperatures similar to those experienced in the beginning of the century.

substantial projected increase in mean annual temperature projected under the two RCP scenarios through the end of the century.

GENERATING THE HINDCAST

The independent variables collected for the analysis are grouped into three categories: periodicity, climate, and time. We utilize these groups in six different combinations to create the models tested in this study. (1) The *periodicity-only* model contains the Fourier transformation, or underlying signal of the observed energy data, as the only input variable. More information on the formulation of the periodicity model can be found in the supporting information, text S2. (2) The *climate-only* model contains 12 input variables: dry bulb temperature, wet bulb temperature, dew point temperature, relative humidity, station pressure, sea pressure, wind speed, precipitation, precipitable water, cloud fraction, cloud opacity, and irradiation. (3) The *periodicity and climate* model consists of 13 variables, including all of the variables from both the periodicity-only and climate-only models. (4) The *periodicity and time* model consists of 47 input variables, including the single variable in the periodicity-only model and all of the categorical time variables (hour of the day [23 variables], day of the week [6], weekday vs weekend [1], month [11], heating vs cooling vs no-heat-no-cool seasons [2], and fiscal year [3]). (5) The *climate and time* model consists of 58 input variables, including all of the variables from the climate-only model and all of the categorical time variables. The last model, (6) the *collective* model, consists of all 59 input variables, including all variables from the periodicity-only and climate-only models, and all categorical time variables. We do not consider a time-only model in the study. Weiss et al. (2022) provides the specific inputs within each of the models.

CROSS-VALIDATED PCR

With the final input variables established, multicollinearity is addressed through cross-validation and PCR (Lins 1985). PCR is a multiple linear regression that uses the principal components (PCs) generated through a PCA of the model inputs. Delorit et al. (2017) explain that PCR is commonly applied in forecasting and hindcasting to reduce both variable dimensionality and multicollinearity, and results in a set of PCs that represent the variance in a set of predictors. First, PCA is conducted where input variables are broken down into their PCs. Next, a leave-one-out cross-validated hindcast is undertaken across the entire dataset to produce a less biased, deterministic prediction of expected energy consumption for WPAFB. Because this form of cross-validation removes the time-step being predicted, the percentage of variance explained by the model will generally decrease. A leave-one-out cross-validation was chosen over k-folds or data-split methods to limit bias in the model, though it does increase computation time, significantly. Furthermore, Jolliffe's rule is applied as a PC retention and dimensionality reduction technique (Jolliffe 1972). Only the most influential PCs for the prediction model are retained. The coefficient of variation falls as fewer PCs are retained for the regression. The cumulative effect of the cross-validated PCR is an unbiased and conservative variance explained estimate.

STATISTICAL CORRECTION

Statistical bias correction, also known as quantile mapping, is prevalent in climate forecast modeling (Cannon et al. 2015; Maraun 2013; Ringard et al. 2017). By comparing the fit of the regressed models to the observed data, statistical model bias can be identified and corrected. Statistical correction methods account for consistent bias across a model. To correct the models, the distribution type of the observed energy data is identified. Using the associated distribution parameters, the distribution of the predictions is matched to the distribution of the observations using quantile mapping. The resultant outputs are the final deterministic prediction models.

The observed energy data follows a bimodal normal distribution that necessitates a uniquely tailored statistical correction process. Normal distribution parameters from both distribution 'modes' are collected to perform the statistical correction. This method requires both the observed and modeled data to be split at a calibrated point while maintaining time-step position indexing. Each 'half' of the modeled data is corrected based on the corresponding 'half' of the observed data, and the two 'halves' of the model are reassembled to produce the statistically corrected model.

VALIDATION METRICS

Deterministic model performance is illustrated using mean absolute percent error (MAPE). MAPE is commonly used in energy prediction research and established thresholds are used to determine the skill of prediction models (Adedeji et al. 2019; Apadula et al. 2012; Capuno et al. 2017; De Felice et al. 2015; Moral-Carcedo and Pérez-García 2019; de Oliveira and Cyrino Oliveira 2018; Panda et al. 2017; Son and Kim 2017). When utilizing MAPE, a score below 20 signifies a prediction model of 'good' quality. If the MAPE falls below 10, the forecast model is said to be of 'excellent' quality (Lewis 1982).

Uncertainty is incorporated into the finalized models through prediction ensemble generation. Ensembles are used to calculate ranked probability skill score (RPSS), which is a metric of probabilistic, or categorical, performance and is meant to account for uncertainty in the deterministic models' outputs. First, a reference climatology is established by separating the distribution of observed data into categories based on the characteristics of the distribution. This becomes the standard against which the prediction ensembles are tested. Climatology is scored based on the percentage of observed data points that fall within each category, while the prediction model is scored based on the number of ensemble predictions that fall in the same category as the observed data.

LEAN MODEL COMPILATION

Once the skill for each of the six deterministic models is evaluated, a lean model is assembled using only the most dominant input variables from the most skillful model. Dominance is determined by correlating the model's retained PCs with the original input variables. The following steps were followed to identify the input variables with the greatest signal:

(a) Isolate the top two-thirds of retained PCs. This decision is arbitrary, but serves to illustrate that an energy manager could down-select to the number of PCs desired based on data availability.
(b) Select input variables from each PC with the absolute value of correlation coefficients greater than 0.30. This is done as 0.30 is widely regarded as "moderate" correlation.
(c) Retained input variables are those that occur most often in the remaining PCs.

The new model is then redeveloped through cross-validated PCR, statistical correction, ensemble generation, and skill analysis. The model's statistical performance is then compared to the initial models to determine the effect of including fewer variables (i.e., the most important variables) from the larger set of input variables. We recognize that the threshold for inclusion of PCs is arbitrary and a different threshold could impact the overall selection of parameters. However, the intent of this exercise is to evaluate whether model predictive skill is retained with a subset of relevant PCs.

A second lean model is created using only daily temperature averages and categorical time inputs to mimic some of the more readily available data from climate projections. The use of this hindcasts serves to validate the use of a forecast model based on these factors.

FORECAST MODEL CALIBRATION AND VALIDATION

Using the lean model developed by down selecting PCs, a forecast of electric consumption is generated. However, the model developed is potentially conservative as it assumes use patterns of the community are unchanging, i.e., the Air Force installation does not grow or change with time. Additionally, it does not consider any efficiency upgrades that would significantly alter overall demand. These assumptions are necessary but could be calibrated for future models based on input from energy managers and new policy. For example, the 2022 Inflation Reduction Act in the United States includes incentives for electrifying facilities, specifically residences. These changes, at a campus scale, would increase the winter electric demands through heat pumps, impacting the determined relationship between temperature and consumption patterns.

Forecasts of daily electric consumption are generated by applying CMIP5 forecasted maximum and minimum temperature inputs and categorical time inputs to the coefficients generated by the cross-validated electric consumption prediction model. The electric consumption forecasts are aggregated to total monthly consumption, and placed in decadal categories (2020–2030, 2030–2040, etc.) to present a range of possible yearly energy consumption values for each year within a decade. Trends in monthly energy consumption are then compared across the century. One-way ANOVA tests are used to determine the significance of monthly energy consumption changes between the first decade and each subsequent decade to highlight when, during the century, monthly and seasonal trends diverge from current use behaviors.

RESULTS

HINDCAST MODEL RESULTS

Deterministically, the top-performing models are the collective and climate and time models, as each produces an explained variance of 0.73 (r^2). Probabilistically, the significance of incorporating statistical correction into model development is manifested as a 9% average increase in RPSS across all models, with the exception of the periodicity-only model (1% improvement), which is likely due to the fact it is extracted from the observed data.

Dimensionality reduction compares the number of retained PCs to the initial number of PCs for each model. Specifically, it is the ratio of the difference between the initial number of PCs and the final number of retained PCs in a model to the final number of retained PCs in a model expressed as a percentage. Reducing dimensionality in a model is important because it reduces the complexity of the model and highlights what input variables are not necessary to produce the model In other words, it narrows the scope of input variables that energy managers must collect and input into a model.

The performance metrics identify that some models do show particularly encouraging skill. All models, except for the periodicity only model, produce MAPE scores consistent with 'excellent' prediction/forecast model candidates (<10); the periodicity-only model is considered 'good'. A total of approximately 100 ensembles for each model formulation were computed to achieve RPSS convergence (<0.01 score deviation). Clearly, higher RPSSs stem from models with larger deterministic variance explained.

Periodicity provides predictive power only when coupled with categorical time variables. When paired with the climate variables, periodicity provides a slight improvement over the climate-only model, and in the collective model, periodicity adds little improvement when compared to the climate and the time model. In direct comparison, the climate-only model performs substantially better than the periodicity-only model. These combined results suggest that the climate inputs provide largely the same information that the periodicity does, but the climate inputs also provide more additional information. As such, for this research, models with climate inputs are favored over those with the periodicity.

Because the climate and time model was the least complex and highest-performing model, it was used to create two lean models consisting of only those input variables with the most dominant signals. After cross-validated PCR, 42 of 58 PCs were retained. Correlating PCs to the specific input variables determined that the inputs with the most dominant signals include the three temperature variables (dew point, dry bulb, and wet bulb) and the time variables weekday/weekend, January, February, June, Sunday, Friday, 1100 h, 1400 h, 1500 h, 1600 h, and 2300 h.

Contingency tables are leveraged to test the categorical performance of the top-performing deterministic model (climate and time). Hits and misses are expressed as a percentage of the total number of forecasts in each climatological category. Climatological categories were defined by the histogram of energy consumption, using local minima and maxima as the cutoffs for categorical definition. A breakdown of these categories and their thresholds are provided in Weiss et al. (2022) in the supporting information. In the contingency tables, hits appear along the diagonal from top left to bottom right. It follows that misses appear as a divergence from the diagonal. The hit scores align closely to model RPSS; however, the extreme miss score is new information, and represents cases when the prediction was for low energy consumption, but the actual energy consumption was high, and vice versa. The hit score of the best performing deterministic model (climate and time model) is 58.6%, and the extreme miss score is 7.9%. Additionally, the hit score for the highest- and lowest-use categories is 72%. While the overall hit score is unimpressive, the model's performance in the extremes is encouraging. If the energy manager's goal is to avoid extreme misses, and maximize skill in predicting extreme use times, then the model should be preferred over a climatological analog.

Being that the high and low categories are likely to be of greatest importance to energy managers, the contingency tables for the median predictions can also be readapted to consolidate the middle two regions (mid-low and mid-high) to a single category because specificity in these regions may not be necessary or important to energy managers. The result is an increased hit score of 67.7% and a decreased extreme miss score of 0.17%. The increase can be attributed to the higher accuracy in the new 'middle' region due to the consolidation of the mid-high and mid-low regions, and the decrease in extreme misses is attributed to the fewer opportunities for values to fall in extreme miss categories.

LEAN MODEL RESULTS

Two lean models are generated with different combinations of input variables to specifically analyze the effect of including categorical time variable types rather than only including specific time variables. Lean model A consists of 44 input variables, including all of the temperature variables (dew point temperature, dry bulb tempera- ture, and wet bulb temperature) and only the most impactful time variable types (hour of the day [23], day of the week [6], weekday vs weekend [1], and month [11]). For example, since several specific hour-of-the-day variables are noted as being impact- ful, the entire set of hour-of-the-day variables were included in the model. Lean model B further down selects the data to match the daily climate data readily avail- able within CMIP5 projects. The model consists of daily temperature and categorical time variables, including day of the week and month.

Lean model A maintains higher performance results compared to the six original models, while lean model B experiences larger drops in performance. This result occurs because lean model A contains more total input variables than lean model B. However, lean model B still outperforms three of the six original models (periodic- ity-only, climate-only, and climate and periodicity) and performs similarly to the periodicity and time model.

FORECAST MODEL RESULTS

The previous analyses focused on hindcasts of the electric consumption for WPAFB. Through the identification of the primary driving factors of electric consumption in the hindcast models, we built an electric demand forecast using climate change ensembles. The 'excellent' MAPE score (<10) provides the basis for the forecasting analysis. We utilize climate ensemble projections for two representative concentration pathways (RCP). RCPs consider a wide array of mitigation efforts for anthropogenic climate change. RCP4.5 is an intermediate scenario with emissions peaking around 2040 before declining. RCP8.5 is a scenario in which emissions continue to increase through the end of the century.

Applying the model in a forecast mode, it is revealed that electric consumption will increase by the end of the century for both RCP cases, though RCP8.5 electric consumption increases more aggressively beginning around year 2065. Between years 2020 and 2040, there is no substantial increase in electric consumption, nor is there a significant difference in consumption predictions for the RCPs. This result could be attributed to the recent observed decreases in temperature since 2017 and milder maximum and minimum temperatures projected for the first two decades following the year 2020. By applying a linear fit to both RCP forecasts, it appears that electric consumption could increase by 0.80 GWh per year for RCP4.5 and 2.14 GWh per year for RCP8.5 through the end of the century. As such, any goals for meeting net-zero energy policies must account for these approximate annual demand changes. Energy managers planning to meet mitigation strategies must build infrastructure and budgets appropriately to meet these projections.

Several months, primarily during the boreal winter and spring, are likely to experience either consistent or falling electric consumption through the year 2040 in both RCP scenarios. By the end of the century, the electric consumption across all months will likely meet or surpass 2020–2030 energy consumption totals for both RCP cases. Spring, summer, and fall months achieve greater energy consumption under RCP8.5, with higher degrees of significance, much sooner than RCP4.5. Again, while this general result is expected, the onset of significantly elevated electric consumption values, and subsequent costs, was unknown until this point. Also, RCP8.5-informed forecasts produce significant increases in winter energy consumption as early as the decade 2040–2050, while RCP4.5 results show decreases in winter energy consumption in this same period. Additionally, RCP8.5 monthly energy consumption exhibits a higher degree of inter-annual variability than RCP4.5, which could mean more uncertainty in forecasted results or less stability in annual consumption. Less stability in consumption would make the task of budgeting on the part of energy managers difficult. Overall, there is high confidence that summer and adjacent seasons' energy consumption will increase earlier in the century, while the timeframe for increases of winter energy consumption may be variable.

DISCUSSIONS

LIMITATIONS OF THE RESEARCH

This research is limited in that the models were calibrated to the singular location of Dayton, OH. Future research must be conducted to evaluate the skill of such models across a span of varying climate regions to validate its adaptability. This is particularly important because the aspects of climate that impact energy use are likely to vary. Therefore, exhaustive inclusion of climate input types should be favored in initial model development to identify which are most impactful in the PCA. Additionally, a large and potentially conservative number of PCs for the models with larger input sets were retained using Jolliffe's rule. Adopting other rules (e.g. Kaiser's rule) could further narrow the retained PC count of larger input sets.

Additionally, the analysis is limited in its exploration of only six models in the initial formulation. A more robust iterative approach for evaluating independent variables might be appropriate to better inform future lean models, such as those in Galelli and Castelletti (2013) and Galelli et al. (2014). However, these iterations should also include weights toward accessibility and ease of access of data for energy managers and other decision-makers.

MODEL PREDICTIVE STRENGTH

The results demonstrate that skillful predictions of hourly campus-wide energy consumption can be achieved using statistical models informed with mixtures of continuous climate and categorical time variables. Moreover, models can be created with techniques (PCR, cross-validation, and statistical correction) that minimize bias and reduce dimensionality. Furthermore, using uncertainty in deterministic predictions, a model's probabilistic skill can be determined. The skill of the proposed framework and use of open-access data suggests that energy and facility managers could be well-positioned to create their own models. The correlation between the regressed PCs illustrates that temperature and time variables are the most useful in explaining hourly energy consumption. Energy consumption patterns were used to decide which categorical time variables to include, while the temperature data was obtained from an open-access NOAA database. Both of these variable types require limited effort to obtain. However, as the comparison of the climate and time model and lean model B shows, there is a significant tradeoff between reducing dimensionality and maintaining skill.

Though overall predictive strength is important, accuracy at the highest and lowest energy use periods is perhaps of greatest importance to energy managers, who must make operational decisions (e.g. load shedding), and make equipment and policy recommendations to decision-makers. For example, predicting peak energy consumption can inform energy managers of when high-demand generators (i.e., those generators only used to compensate for peak energy periods), should be utilized or if energy infrastructure needs expansion to support increased demand. Additionally, hourly predictions facilitate greater integration with renewables that are predictable but vary in availability throughout a day. Predicting low-energy periods accurately can inform seasonal decisions to override heating and cooling systems when environmental conditions are mild (e.g. spring and fall) (Delorit et al. 2020). These predictions offer opportunities both in the short term for annual budgeting but also in the long term to develop energy transition pathways to meet net-zero goals for the federal government.

IMPACT FOR CARBON FOOTPRINT

The results of the century forecast model illustrate that starting in 2040 there will be significant annual and seasonal deviations from the historic average of electric consumption. For RCP4.5 and RCP8.5 scenarios, an additional demand of 0.80 GWh and 2.14 GWh per year incurs a significant burden not just in terms of cost, but also with respect to reducing emissions. Previous assessments of Dayton, Ohio's greenhouse gas emission intensity ranged from 370 to 1710 kg CO_2/MWh with an average of 684 kg CO_2/MWh depending on the accounting method (Siddik et al. 2020). Utilizing these emissions intensities, this equates to an additional 547 000 kg CO_2 or 1460 000 kg CO_2 for RCP4.5 and RCP8.5, respectively, each year. Current emissions from electric consumption at WPAFB total approximately 547 million kg CO_{2c} for the year 2020 and could rise to nearly 700 million kg CO_2 by the end of the century for RCP8.5, considering no change in the existing generation mix. Therefore, to meet EO 14057, WPAFB will need to plan for significant additional electric demand by the 2050 deadline for net-zero energy, approximately an additional 50 GWh of demand. Without using appropriate forecasting tools to assess increased electric consumption, the installation could miss these important climate targets.

IMPLICATIONS FOR POLICY AND ENERGY MANAGERS

The forecasting results are consistent with similar works in this field of study. The energy consumption forecasts developed in this research show a constant, and even a slight decline, in energy consumption approaching the middle of the century. The work of Zhou et al. (2014) and Emodi et al. (2018) capture this phenomenon. For example, Zhou et al. finds that a slight decline in heating demand and a slight decline in cooling demand occur in the first half of the 21st century for the state of Ohio. Since the primary facility energy drain related to climate is heating and cooling, Zhou et al. appears to explain a large part of what is observed with total energy consumption in the research herein. In contrast to existing studies, the forecasts developed in this research aid in analyzing century-long trends in energy consumption at the campus level using campus-level energy data. This research suggests that energy managers and campus leaders must be prepared for energy consumption, and subsequent costs, to increase over the course of the century. Long-term consumption and cost forecasts, consistent with the type produced here, provide valuable information to mitigate climate impacts and meet renewable and net-zero energy targets.

Energy managers are generally not modelers, and thus tools that are informed with readily available data are likely to be favored. Data accessibility, computational power, modeling ability, and time availability could be factors in model construction. Though models with climate variables tend to outperform less data-intensive constructs, managers may favor a periodicity-based model as it only requires the energy consumption data itself. Ultimately, both approaches are viable and can produce skillful models. The information provided by these forecasts enable campus energy managers to understand the magnitude and time frame where energy consumption and costs could significantly escalate. Campus managers can decide what risk-mitigation strategies to implement, decide when to implement them, and economically justify their investment decisions. To mitigate and overcome cost increases, energy managers should consider policy (near-term) and infrastructure (long-term) adaptations to overcome increases. This model framework can then be used as a tool to justify the economic benefits of infrastructure and renewable resource investments.

CONCLUSIONS

EO 14057 directs federal agencies to achieve 100 percent carbon-free electricity on a net annual basis by 2030 and net-zero emissions by 2050. Using WPAFB, OH as a case study, we demonstrate that electric demand is expected to significantly rise throughout the remainder of the century. To meet these targets, federal agencies need robust predictive practices when building out electric infrastructure to meet not just current demands but also future projections. In this study, we present several options for creating robust forecasts of electric demand at a campus or installation scale. These type of analyses are integral to facilitate data-driven decisions focused on climate-informed infrastructure. The methodology developed here provides a flexible framework that can be adapted to any number of continuous or categorical independent variables, utilizes open-source data, and extensively accounts for modeling bias. Each model configuration presented performs skillfully in those areas most important for decision-making and policy development. Energy managers and policy-makers should look to climate projections and skillful forecasts to create robust predictions for net-zero and carbon-free infrastructure investments. For additional details of this research, including data availability and sharing, figures, tables, and statistical results, readers are directed to the original paper (Weiss et al. 2022).

REFERENCES

Adedeji, P. A., S. Akinlabi, N. Madushele, and O. Olatunji (2019), "Neuro-Fuzzy Mid-Term Forecasting of Electricity Consumption Using Meteorological Data," *IOP Conference Series: Earth and Environmental Science*, Vol. 331, p. 012017.

Al-bayaty, H., T. Mohammed, A. Ghareeb, and W. Wang (2019), "City Scale Energy Demand Forecasting Using Machine Learning Based Models: A Comparative Study," *Proceedings of the Second International Conference on Data Science, E-Learning and Information Systems, DATA '19*, Association for Computing Machinery, Dubai, United Arab Emirates, 1–9.

Amato, A. D., M. Ruth, P. Kirshen, and J. Horwitz (2005), "Regional Energy Demand Responses to Climate Change: Methodology and Application to the Commonwealth of Massachusetts," *Climatic Change*, Vol. 71, No. 1–2, pp. 175–201.

Apadula, F., A. Bassini, A. Elli, and S. Scapin (2012), "Relationships between Meteorological Variables and Monthly Electricity Demand," *Applied Energy*, Vol. 98, pp. 346–356.

Cannon, A. J., S. R. Sobie, and T. Q. Murdock (2015), "Bias Correction of GCM Precipitation by Quantile Mapping: How Well Do Methods Preserve Changes in Quantiles and Extremes?" *Journal of Climate, American Meteorological Society*, Vol. 28, No. 17, pp. 6938–6959.

Capuno, M., J.-S. Kim, and H. Song (2017), "Very Short-Term Load Forecasting Using Hybrid Algebraic Prediction and Support Vector Regression," *Mathematical Problems in Engineering*, Vol. 2017, pp. 1–9.

Chandramowli, S. N. and F. A. Felder (2014), "Impact of Climate Change on Electricity Systems and Markets – A Review of Models and Forecasts," *Sustainable Energy Technologies and Assessments*, Vol. 5, pp. 62–74.

Chowdhury, A. K., T. D. Dang, H. T. Nguyen, R. Koh, and S. Galelli (2021), "The Greater Mekong's Climate-Water-Energy Nexus: How ENSO-Triggered Regional Droughts Affect Power Supply and CO_2 Emissions," *Earth's Future*, Vol. 9, No. 3, e2020EF001814.

Chukhrova, N. and A. Johannssen (2019), "Fuzzy Regression Analysis: Systematic Review and Bibliography," *Applied Soft Computing*, Vol. 84, p. 105708.

Cook, M. A., C. W. King, F. T. Davidson, and M. E. Webber (2015), "Assessing the Impacts of Droughts and Heat Waves at Thermoelectric Power Plants in the United States Using Integrated Regression, Thermodynamic, and Climate Models," *Energy Reports*, Vol. 1, pp. 193–203.

De Felice, M., A. Alessandri, and F. Catalano (2015), "Seasonal Climate Forecasts for Medium-Term Electricity Demand Forecasting," *Applied Energy*, Vol. 137, pp. 435–444.

de Oliveira, E. M. and F. L. Cyrino Oliveira (2018), "Forecasting Mid-Long Term Electric Energy Consumption Through Bagging ARIMA and Exponential Smoothing Methods," *Energy*, Vol. 144, pp. 776–788.

De Rosa, M., V. Bianco, F. Scarpa, and L. A. Tagliafico (2014), "Heating and Cooling Building Energy Demand Evaluation; A Simplified Model and a Modified Degree Days Approach," *Applied Energy*, Vol. 128, pp. 217–229.

Delorit, J., E. C. G. Ortuya, and P. Block (2017), "Evaluation of Model-Based Seasonal Streamflow and Water Allocation Forecasts for the Elqui Valley, Chile," *Hydrology and Earth System Sciences; Katlenburg-Lindau*, Vol. 21, No. 9, pp. 4711–4725.

Delorit, J. D., S. J. Schuldt, and C. M. Chini (2020), "Evaluating an Adaptive Management Strategy for Organizational Energy Use Under Climate Uncertainty," *Energy Policy*, Vol. 142, p. 111547.

Dirks, J. A., W. J. Gorrissen, J. H. Hathaway, D. C. Skorski, M. J. Scott, T. C. Pulsipher, M. Huang, Y. Liu, and J. S. Rice (2015), "Impacts of Climate Change on Energy Consumption and Peak Demand in Buildings: A Detailed Regional Approach," *Energy*, Vol. 79, pp. 20–32.

Emodi, N. V., T. Chaiechi, and A. R. A. Beg (2018), "The Impact of Climate Change on Electricity Demand in Australia," *Energy & Environment*, Vol. 29, No. 7, pp. 1263–1297.

Fan, J.-L., B.-J. Tang, H. Yu, Y.-B. Hou, and Y.-M. Wei (2015), "Impact of Climatic Factors on Monthly Electricity Consumption of China's Sectors," *Natural Hazards*, Vol. 75, No. 2, pp. 2027–2037.

Galelli, S. and A. Castelletti (2013), "Tree-Based Iterative Input Variable Selection for Hydrological Modeling," *Water Resources Research*, Vol. 49, No. 7, pp. 4295–4310.

Galelli, Stefano, et al., (2014), "An Evaluation Framework for Input Variable Selection Algorithms for Environmental Data-Driven Models," *Environmental Modelling & Software*, Vol. 62, pp. 33–51.

Graafland, C. E., J. M. Gutiérrez, J. M. López, D. Pazó, and M. A. Rodríguez (2020), "The Probabilistic Backbone of Data-Driven Complex Networks: An Example in Climate," *Scientific Reports, Nature Publishing Group*, Vol. 10, No. 1, p. 11484.

Ismail, N. and S. Abdullah (2016), "Principal Component Regression with Artificial Neural Network to Improve Prediction of Electricity Demand," *International Arab Journal of Information Technology*, Vol. 13, No. 1A, pp. 196–202.

Jolliffe, I. T. (1972), "Discarding Variables in a Principal Component Analysis. I: Artificial Data," *Journal of the Royal Statistical Society: Series C (Applied Statistics)*, Vol. 21, No. 2, pp. 160–173.

Kim, A. A., Y. Sunitiyoso, and L. A. Medal (2019), "Understanding Facility Management Decision Making for Energy Efficiency Efforts for Buildings at a Higher Education Institution," *Energy and Buildings*, Vol. 199, pp. 197–215.

Lewis, C. D. (1982), *Industrial and Business Forecasting Methods: A Practical Guide to Exponential Smoothing and Curve Fitting*. Butterworth-Heinemann, London.

Lins, H. F. (1985), "Interannual Streamflow Variability in the United States Based on Principal Components," *Water Resources Research*, Vol. 21, No. 5, pp. 691–701.

Mansur, E. T., R. Mendelsohn, and W. Morrison (2008), "Climate Change Adaptation: A Study of Fuel Choice and Consumption in the US Energy Sector," *Journal of Environmental Economics and Management*, Vol. 55, No. 2, pp. 175–193.

Maraun, D. (2013), "Bias Correction, Quantile Mapping, and Downscaling: Revisiting the Inflation Issue." *Journal of Climate, American Meteorological Society*, Vol. 26, No. 6, pp. 2137–2143.

Maurer, E. P., L. Brekke, T. Pruitt, and P. B. Duffy (2007), "Fine-Resolution Climate Projections Enhance Regional Climate Change Impact Studies," *Eos Transactions American Geophysical Union*, Vol. 88, No. 47, p. 504.

McCall, James Colorado, and Jordan Macknick (2016), *Water-Related Power Plant Curtailments: An Overview of Incidents and Contributing Factors*. National Renewable Energy Lab. (NREL), Golden.

Moral-Carcedo, J., and J. Pérez-García (2019), "Time of Day Effects of Temperature and Daylight on Short Term Electricity Load," *Energy*, Vol. 174, pp. 169–183.

Mukherjee, S., and R. Nateghi (2017), "Climate Sensitivity of End-Use Electricity Consumption in the Built Environment: An Application to the State of Florida, United States," *Energy*, Vol. 128, pp. 688–700.

Mukherjee, S., C. R. Vineeth, and R. Nateghi (2019), "Evaluating Regional Climate-Electricity Demand Nexus: A Composite Bayesian Predictive Framework," *Applied Energy*, Vol. 235, pp. 1561–1582.

Panda, S. K., S. N. Mohanty, and A. K. Jagadev (2017), "Long Term Electrical Load Forecasting: An Empirical Study across Techniques and Domains," *Indian Journal of Science and Technology*, Vol. 10, No. 26, pp. 1–16.

Psiloglou, B. E., C. Giannakopoulos, S. Majithia, and M. Petrakis (2009), "Factors Affecting Electricity Demand in Athens, Greece and London, UK: A Comparative Assessment," *Energy*, Vol. 34, No. 11, pp. 1855–1863.

Reclamation (2013), "Downscaled CMIP3 and CMIP5 Climate and Hydrology Projections: Release of Downscaled CMIP5 Climate Projections, Comparison with preceding Information, and Summary of User Needs", prepared by the U.S. Department of the Interior, Bureau of Reclamation, Technical Services Center, Denver, Colorado, 47 pp.

Ringard, J., F. Seyler, and L. Linguet (2017), "A Quantile Mapping Bias Correction Method Based on Hydroclimatic Classification of the Guiana Shield," *Sensors*, Vol. 17, No. 6, pp. 1413–1422.

Siddik, M. A. B., C. M. Chini, and L. Marston (2020), "Water and Carbon Footprints of Electricity Are Sensitive to Geographical Attribution Methods," *Environmental Science & Technology*, Vol. 54, No. 12, pp. 7533–7541.

Solcast (2019), "Global Solar Irradiance Data and PV System Power Output Data", https://solcast.com/solar-radiation-map/ (accessed April 22, 2023).

Son, H. and C. Kim (2017), "Short-Term Forecasting of Electricity Demand for the Residential Sector Using Weather and Social Variables," *Resources, Conservation and Recycling*, Vol. 123, pp. 200–207.

Theusme, C., L. Avendaño-Reyes, U. Macías-Cruz, A. Correa-Calderón, R. O. García-Cueto, M. Mellado, L. Vargas-Villamil, and A. Vicente-Pérez (2021), "Climate Change Vulnerability of Confined Livestock Systems Predicted Using Bioclimatic Indexes in an Arid Region of México," *Science of The Total Environment*, Vol. 751, p. 141779.

Weiss, Scott C., Justin D. Delorit, and Christopher M. Chini (2022), "Energy Forecasting to Benchmark for Federal Net-Zero Objectives Under Climate Uncertainty," *Environmental Research: Infrastructure and Sustainability*, Vol. 2, No. 4, pp. 1–14. https://doi.org/10.1088/2634-4505/ac9712.

Wenz, L., A. Levermann, and M. Auffhammer (2017), "North–South Polarization of European Electricity Consumption Under Future Warming," *Proceedings of the National Academy of Sciences*, Vol. 114, No. 38, pp. E7910–E7918.

Xie, J. and T. Hong (2017), "Wind Speed for Load Forecasting Models," *Sustainability*, Vol. 9, No. 5, p. 795.

Zeng, P., X. Sun, and D. J. Farnham (2020), "Skillful Statistical Models to Predict Seasonal Wind Speed and Solar Radiation in a Yangtze River Estuary Case Study," *Scientific Reports, Nature Publishing Group*, Vol. 10, No. 1, p. 8597.

Zhou, Y., L. Clarke, J. Eom, P. Kyle, P. Patel, S. H. Kim, J. Dirks, E. Jensen, Y. Liu, J. Rice, L. Schmidt, and T. Seiple (2014), "Modeling the Effect of Climate Change on U.S. State-Level Buildings Energy Demands in an Integrated Assessment Framework," *Applied Energy*, Vol. 113, pp. 1077–1088.

Appendix A
System Design, Improvement, and Management Tools Relevant for Environmental and Climate Change Actions

5S
5Y
7 Wastes
8D problem solving

Action Plan
Activity Network Diagram
Affinity Diagram
ANOVA
Attribute Control Charts
Auditing

Balanced Scorecards
BCS
Benchmarking
Block Diagram
Brainstorming

Capability study
Cause and Effect
CCC
Check List
Check Sheets
Chi Square
Consensus Agreement
Control Chart
Cost of Quality
Critical Path Method (CPM)
Critical Path Drag
Critical Resource Diagramming
CTQ tree

Customer Focus Surveys
Customer Needs Analysis
Cycle time studies

Data Analysis Explained
Data Analysis in Microsoft Excel
Data Analysis Techniques
Data Collection Form
Data Envelopment
Database Deployment
Decision Matrix
DEJI Systems Model
Design of Experiments
DFSS
DMAIC
Dot Plot

Earned Value Analysis
Effective Team Meetings
Error Proofing

Fishbone Diagram
Flowchart
FMEA
Force Field Analysis

Gage R&R
Gantt Charts
Graphical Analysis

Histogram

Interaction Process Analysis
Interrelationship Diagram
Interview
Is/Is-not Matrix

Kaizen
Kanban

Linear Regression
LTV

Matrix Diagram
Multivoting

Nominal Group Technique
Normal Distribution

OEE
Organized Teams

Pareto Charts
Paynter Chart
Payoff Matrix
PERT/CPM
PICK chart
Picture Boarding
Pie Charts
Planning For Influence Chart
Precedence Diagramming
Prioritization Matrix
Problem Definition Checklist
Process Decision Program Chart
Process Mapping
Process Standardization
Process Streamlining
Process Summary Worksheet
Project Charter
Project Selection Checklist

Q-MAP
QMS Review
Quality Costs
Questionnaire

Reduce Variation
Regression
Relationship Chart
Responsibility Chart
Run Chart

Sampling
Scatter Diagram
Scenario Analysis
Set Specifications
Simulation Modeling
SIPOC
SMED
Solution Vision Matrix
Spaghetti Diagram
Stakeholder Chart
Statistical Process Control
Statistical Sampling
Streamlining

204

Surveys
Surveys
SWOT Analysis

Takt Time
Target Shifting
Team Dynamics
Team Progress Model
Theory of Constraints
Training Analysis
Tree Diagram
Triple C Model
TRIZ

Value Stream Mapping

Waste Walk

Appendix B
Global Climate Research-Oriented Academies

Academies around the world provide a foundation for international collaboration on intellectual grounds, which facilitates effective multidisciplinary research-management strategies for responding to climate change. Although the most-recognized academies are technically oriented (e.g., engineering), the variety and diversity of academies can cover non-technical disciplines that fit the global systems premise of this book. The International Council of Academies of Engineering and Technological Sciences (CAETS) is one global coalition of technical groups dedicated to the advancement of research in various disciplines, including climate change. As a ready reference, the member academies of CAETS are provided in this Appendix.

MEMBER ACADEMIES OF CAETS

Source: https://www.newcaets.org/membership-2/about-member-academies/, Accessed September 10, 2021

ARGENTINA

Academia Nacional de Ingenieria (ANI)

- Founded: 1970
- Elected to CAETS: 1999
- Website: www.acading.org.ar

AUSTRALIA

Australian Academy of Technology and Engineering (ATSE)

- Founded: 1976
- Founding member of CAETS: 1978
- Website: www.atse.org.au

BELGIUM

Royal Belgian Academy Council of Applied Sciences (BACAS)

- Founded: 1987
- Elected to CAETS: 1990
- Website: www.kvab.be

CANADA

Canadian Academy of Engineering (CAE)

- Founded: 1987
- Elected to CAETS: 1991
- Website: www.cae-acg.ca

CHINA

Chinese Academy of Engineering (CAE)

- Founded: 1994
- Elected to CAETS: 1997
- Website: www.cae.cn

CROATIA

Croatian Academy of Engineering (HATZ)

- Founded: 1993
- Elected to CAETS: 2000
- Website: www.hatz.hr

CZECH REPUBLIC

Engineering Academy of the Czech Republic (EACR)

- Founded: 1995
- Elected to CAETS: 1999
- Website: www.eacr.cz

DENMARK

Danish Academy of Technical Sciences (ATV)

- Founded: 1937
- Elected to CAETS: 1987
- Website: www.atv.dk

FINLAND

Council of Finnish Academies (CoFA)

- Established: 2018
- Elected to CAETS: 1989 (FACTE)
- Website: www.academies.fi

FRANCE

National Academy of Technologies of France (NATF)

- Founded: 1982 (CADAS); 2000 (NATF)
- Elected to CAETS: 1989
- Website: www.academie-technologies.fr

GERMANY

National Academy of Science and Engineering (acatech)

- Founded: 1997
- Elected to CAETS: 2005
- Website: www.acatech.de

HUNGARY

Hungarian Academy of Engineering (HAE)

- Founded: 1990 (MMA)
- Elected to CAETS: 1995
- Website: www.mernokakademia.hu

INDIA

Indian National Academy of Engineering (INAE)

- Founded: 1987
- Elected to CAETS: 1999
- Website: www.inae.in

REPUBLIC OF IRELAND AND NORTHERN IRELAND*

Irish Academy of Engineering (IAE)

- Founded: 1997
- Elected to CAETS: 2020
- Website: www.iae.ie
- *IAE is an all-island body*

JAPAN

Engineering Academy of Japan (EAJ)

- Founded: 1987
- Elected to CAETS: 1990
- Website: www.eaj.or.jp

Appendix B

KOREA

National Academy of Engineering of Korea (NAEK)

- Founded: 1995
- Elected to CAETS: 2000
- Website: www.naek.or.kr

MEXICO

Academy of Engineering of Mexico (AIM)

- Founded: 1973-Mexican Academy of Engineering; 1974-National Academy of Engineering Mexico; 2002-Combined to AIM
- Founding member of CAETS: 1978
- Website: www.ai.org.mx

NETHERLANDS

Netherlands Academy of Technology and Innovation (ACTI.nl)

- Founded: 1986
- Elected to CAETS: 1993
- Website: www.acti-nl.org

NEW ZEALAND

Royal Society Te Aparangi

- Founded: 1867
- Elected to CAETS: 2019
- Website: www.royalsociety.org.nz

NIGERIA

Nigerian Academy of Engineering (NAE)

Founded: 1997
Elected to CAETS: 2019
Website: www.nae.org.ng

NORWAY

Norwegian Academy of Technological Sciences (NTVA)

- Founded: 1955
- Elected to CAETS: 1990
- Website: www.ntva.no

PAKISTAN

Pakistan Academy of Engineering (PAE)

- Founded: 2013
- Elected to CAETS: 2018
- Website: www.pacadengg.org

SERBIA

Academy of Engineering Sciences of Serbia (AESS)

- Founded: 1998
- Elected to CAETS: 2019
- Website: www.ains.rs

SLOVENIA

Slovenian Academy of Engineering (IAS)

- Founded: 1995
- Elected to CAETS: 2000
- Website: www.ias.si

SOUTH AFRICA

South African Academy of Engineering (SAAE)

- Founded: 1991 (Academy of Engineers in South Africa)
- Elected to CAETS: 2009
- Website: www.saae.co.za

SPAIN

Real Academia de Ingenieria (RAI)

- Founded: 1994
- Elected to CAETS: 1999
- Website: www.raing.es

SWEDEN

Royal Swedish Academy of Engineering Sciences (IVA)

- Founded: 1919
- Founding member of CAETS: 1978
- Website: www.iva.se

SWITZERLAND

Swiss Academy of Engineering Sciences (SATW)

- Founded: 1981
- Elected to CAETS: 1988
- Website: www.satw.ch

UNITED KINGDOM

Royal Academy of Engineering (RAEng)

- Founded: 1976
- Founding member of CAETS: 1978
- Website: www.raeng.org.uk

UNITED STATES OF AMERICA

National Academy of Engineering (NAE)

- Founded: 1964
- Founding member of CAETS: 1978
- Website: www.nae.edu

URUGUAY

National Academy of Engineering of Uruguay (ANIU)

- Founded: 1971
- Elected to CAETS: 2000
- Website: www.aniu.org.uy

Index

Pages in *italics* refer figures.

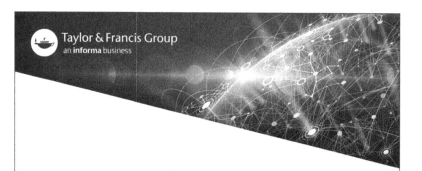

Printed in the United States
by Baker & Taylor Publisher Services